A Cumbria

HERDWICKS

Herdwick Sheep and the English Lake District

Geoff Brown

HERDWICK, noun: a sheep of a hardy mountain breed from the north of England – ORIGIN early 19th century from (now obsolete) herdwick 'pasture ground'... *Oxford Dictionary of English*, 2005, 2nd edition, eds. C. Soanes and A. Stevenson

HERDWICK, (UOR'D.WIK) – The mountain sheep of Cumberland. *Glosssary of the Dialect of Cumberland*, 1879, by William Dickinson, ed. E. W. Prevost

HAYLOFT PUBLISHING LTD

First published by Hayloft 2009
This reprint 2012

Hayloft Publishing Ltd, South Stainmore, Kirkby Stephen,
Cumbria, CA17 4DJ

tel: 017683 41568
email: books@hayloft.eu
web: www.hayloft.eu

Copyright © Geoff Brown, 2009

ISBN 1 904524 65 6 (paperback)
ISBN 1 904524 66 4 (hardback)

CAP data for this title are available from the British Library

Apart from any fair dealing for the purposes of research or private study or criticism or review, as permitted under the Copyright Designs and Patents Act 1988 this publication may only be reproduced, stored or transmitted in any form or by any means with the prior permission in writing of the publishers, or in the case of reprographic reproduction in accordance with the terms of the licences issued by the Copyright Licensing Agency.

Designed, printed and bound in the EU

Papers used by Hayloft are natural, recyclable products made from wood grown in sustainable forest. The manufacturing processes conform to the environmental regulations of the country of origin.

Preface

I have been collecting information, in one way or another, about fell farming in the Lake District of the north of England for many years. From an early age I subjected my uncle, William Rawling Snr. of the Hollins, Ennerdale, to numerous questions about Herdwick sheep, the way they were kept and the people who kept them. For most of the past 25 years I have bred Herdwick sheep myself and I was the secretary of the Herdwick Sheep Breeders' Association between 1990 and 2008. In this capacity I have been involved in the registration of young Herdwick tups in the annual inspections that take place on the main Herdwick farms in September each year. I learnt much in the early years from my journeys round these farms throughout the Lake District with members of the tup inspection panel: notably William Bowes, William Bland, Derwent Tyson, J. V. Gregg, George Birkett, Jack Bland and Joe Folder. In more recent years I have been accompanied by members of the HSBA Executive Committee and have enjoyed their company and the task. I have also learnt a good deal over the years about the policy dimension of fell farming from Andrew Humphries, Will Cockbain and Mervyn Edwards and wish to thank them for sharing their knowledge and insights.

I would like also to thank Louise Rawling of the Hollins and Dorothy Wilkinson of Tilberthwaite for the work they did in collecting photographs as part of the 90th birthday celebrations of the HSBA in 2006, many of which are included here – thanks also to Mrs Rene Walker. Thanks also to Ian Brodie who supplied the cover photograph which perfectly captures Herdwick sheep above Buttermere overlooking Dale Head.

I wish to thank the people who read and commented on versions of the text: William Rawling Snr, William F. Rawling, Angus Winchester, Susan Denyer, Ian Brodie and Jane Thomas. I am particularly grateful to Terry McCormick for commenting in detail on the text and for his suggestions for improving the final section of the book in particular, and to Dawn Robertson for her expertise in preparing this book for publication

I wish finally to thank all members of the Herdwick sheep breeding community for their commitment over the years: without them this book would not have been possible.

Geoff Brown, May 2009

This book is dedicated to the sheep dogs of the Lakeland fells.

Contents

PREFACE   *page*   3

1. ORIGINS AND DEVELOPMENT OF THE HERDWICK BREED
Where do Herdwick Sheep Come From?	7
A Viking Legacy?	8
An evolving Breed?	10
The Fell Dales Association	13
Defining the Breed Type	18
Canon Rawnsley and Herdwick Sheep	23
Forming the Herdwick Sheep Breeders' Association	24
Depression and Difficulties	31
Agreeing Breed Standards	35
Losing Ground	39

2. FELLS AND HEAFS
Heaf-going Sheep, Fell-going Sheep	42
Landlords' Flocks	43
Wool and Wethers	48
Farming the Fells	52
Shepherd's Guides and Shepherds' Meets	56
Shepherding	62
Sheepdogs and Sheepfolds	64
Snows, Diseases and Disasters	67

3. HERDWICKS AND THE LAKE DISTRICT LANDSCAPE
Part of the Landscape	72
The Threat of Afforestation	73
National Property, National Trust and National Park	76
The Tale of Mrs Heelis	87
Lake District Farm Estates	94
Creating the National Trust Fell Farming Estate	95
Significance of Herdwick Sheep Farming	97
The Lake District National Park	100

4. RECENT YEARS: CHALLENGES AND CHANGES
The Rise of the Swaledale	103
Years of Expansion: The Numbers Game	110

The Agri-Environment Challenge	113
Foot and Mouth Disease, 2001	115
English Nature and FMD	117
The National Trust and the Foot and Mouth Crisis	118
Reform of the CAP	122

5. THE WAY FORWARD FOR HERDWICK SHEEP

Recent Issues in Herdwick Sheep Keeping	124
Grazing for Conservation?	125
Farm Animal Genetic Resources	126
Adding Value to Herdwick Products	128
New Markets for Live Sheep?	132

6. HERDWICK COUNTRY

Herdwick Country: A Cultural Landscape	135
The Role of the National Trust	136
An Agenda for the National Park	142
Cultural Landscape: Towards a World Heritage Site?	146

ANNEX

Cumbrian Fell-going Flocks	153
Smaller Cumbrian Flocks	155

NOTES AND REFERENCES

Chapter 1	156
Chapter 2	161
Chapter 3	166
Chapter 4	171
Chapter 5	172
Chapter 6	173

INDEX

People, Practices and Organisations	175
Farms	179
Photographs	181

1: Origins and Development of the Herdwick Breed

"There is a kind of sheep in these mountains called Herdwicks... these sheep lye upon the very tops of the mountains in that season [winter] as well as in summer." James Clarke, A Survey of the Lakes (1787)

"...their flocks rise from an hundred to a thousand... they keep them, in both winter and spring, on the commons." Arthur Young, A Six Month Tour through the North of England (1768)

Where do Herdwick Sheep Come From?

The high fells of the English Lake District are home to the Herdwick sheep. The Herdwick breed of sheep is widely thought to be the hardiest of all Britain's sheep breeds. Herdwick sheep live on the fells for virtually all of the year on England's highest and roughest terrain, an area which also has England's highest rainfall. 'Herdwick Country' essentially consists of the western and central parts of the Lake District National Park. The central and western fells of this area – covering Borrowdale, Buttermere, Ennerdale, Wasdale, Eskdale, the Duddon Valley, Coniston, the Langdales and Grasmere and in the Helvellyn area around Ullswater and Thirlmere – make up England's only mountain area, and are the breed's main stamping grounds. These ancient sheep walks (known as 'heafs') are Herdwick country.

This area in geological terms is largely an area of volcanic and sedimentary rocks and forms the central dome of the Lake District. William Wordsworth, nearly 200 years ago, described the essence of the topography of the area when he described the area around Scafell and Great Gable as being like the hub of a great cartwheel with the valleys radiating from it like spokes of that wheel. He also pointed out that from the centre – say on Esk Hause – a shepherd could descend within a relatively short space of time to one of several dales.

"From a point between Great Gavel and Scawfell, a shepherd would not require more than an hour to descend into any one of eight of the principal vales by which he would be surrounded; and all the others lie (with the exception of Haweswater) at but a small distance."[1]

It is here on England's highest ground with its open sweeps of fell mixed with massive rocks and steep crags, that the breed has been developed by those shepherds and their descendents as a sheep able to withstand that terrain, and to live and reproduce where other breeds of sheep do much less well. Between 100 and 120 farms from the dales that use these fells still keep Herdwick sheep in commercial numbers, with many of them running flocks going back for many years. Herdwick sheep are a central feature in that landscape and they are integral to the culture of the area.

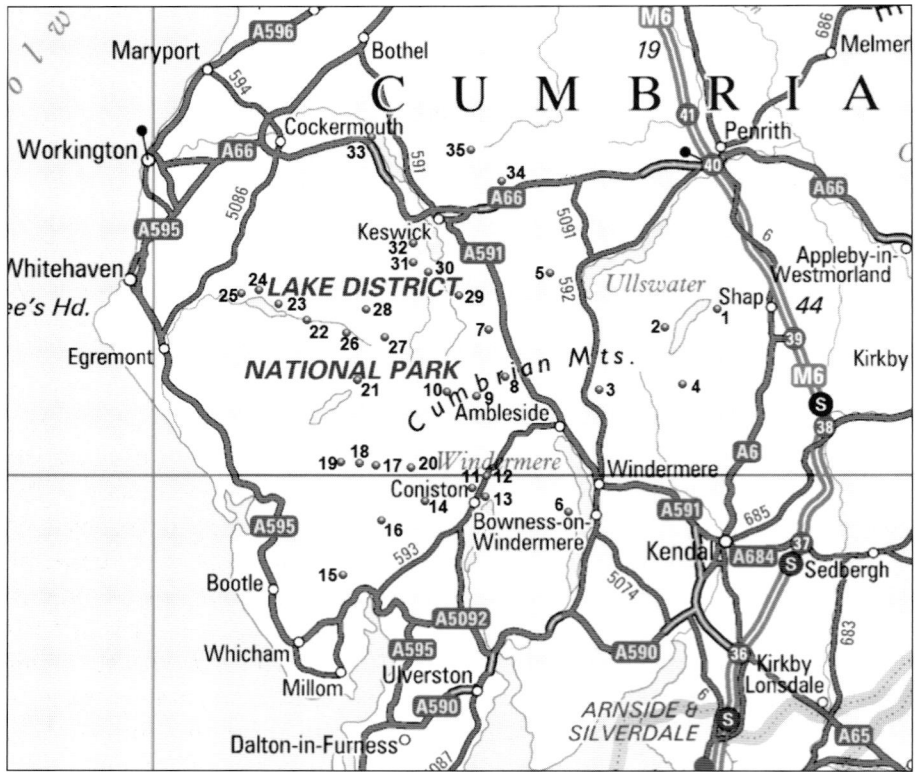

KEY to some of the many places mentioned in the text:
1. Naddle
2. Mardale Dun Bull
3. Troutbeck Park
4. Kentmere Hall
5. Glencoyne
6. Hill Top
7. West Head
8. Brimmer Head
9. Mill Beck
10. Middlefell
11. Tilberthwaite
12. Yew Tree, Coniston
13. Yewdale
14. Turner Hall
15. Fenwick
16. Hazel Head
17. Butterilket
18. Taw House
19. Woolpack, Eskdale
20. Black Hall
21. Middle Row, Wasdale
22. Gillerthwaite
23. Mireside
24. How Hall
25. Hollins
26. Black Sail Huts
27. Seathwaite, Borrowdale
28. Gatesgarth
29. Watendlath
30. Yew Tree, Rosthwaite
31. Nook
32. Skelgill
33. Herdwick View
34. Middle Row, Threlkeld
35. Skiddaw Forest

A Viking Legacy?

Much attention has been given to the origins of Herdwick sheep – with much and oft-repeated speculation and mythology. Some have asserted (and many have repeated it) that they swam ashore from a wrecked Spanish Armada galleon, washed up around Drigg and then were shared out amongst the farmers of the Wasdale area. This can be discounted without any further ado except to note that quite a number of Britain's livestock breeds are reputed to have this dubious background. The other great, again often-repeated, story is that Herdwicks came here with the Vikings. There is possibly just a little more truth in this one: but the reality is complex and uncertain.

The simplified story seems to be that the Vikings began the process of farming the fells seriously and that they brought Herdwick sheep with them in order to do it. There is extremely little evidence for this. The picture seems to be (from archaeological and other evidence) that Vikings came to Cumbria especially from the early ninth century. There is no evidence that the Viking population led much of an independent existence from other people and so it is presumed that they mixed with existing local people. The leading expert on the Viking influence on the Lake District, Professor Ian D.Whyte, points out that not a single settlement site can be attributed to the Norse colonisation period. But in the Lake District, above the level of improved land, foundations of buildings can be found which might either be sixteenth century homesteads, medieval shielings, or just possibly Viking homesteads – evidence perhaps of there being a Norse system whereby livestock were taken to summer grazing in the fells. The existing Anglian people were presumably already exploiting the better land areas both in the lowlands, and in the fells, and there seems to be evidence that the Norse settlers colonised the less fertile areas around these places.

The extent and rapidity of deforestation shows that there may have been a substantial penetration of upper valleys, as the place-name evidence all over the Lake District seems to suggest.[2] Between the ninth century and 1066, there seems to have been a general build up of the native population, which led to an expansion of settlement in the Lake District, and it may well have been that the Scandinavians gave this movement a distinctive character, given their interest in pastoralism. The Vikings may have followed their sheep-keeping practices possibly using sheep that were already here.

They came to Cumbria from the west, via Ireland and the Isle of Man, and it is, of course, possible that they brought some sheep into Cumbria with them. The Reverend T. Ellwood of Torver writing in 1897 concluded that Herdwick sheep were introduced by Viking invaders because of the similarity of Norwegian and local sheep marks. As has been said, however, this is "evidence of the origin of the shepherds rather than of the sheep."[3] The assumption that some people make is that the Norse settlers must have been primarily sheep farmers because of the "supposed Scandinavian origins of the Herdwick breed" has, according to Whyte, yet to be adequately demonstrated.[4]

David Kinsman, who has recently carried out extensive research on primitive sheep breeds, especially the Hebridean (which as he has shown has strong Cumbrian connections in its 'park sheep' version), has pointed out that the first – very primitive – sheep were introduced into Britain about 5,500 years ago with a more advanced type – similar to the Shetland, Orkney, Hebridean and Manx – arriving about 1,500 years later. At the time of the arrival of the Vikings into Britain, about a thousand years ago, they would have encountered sheep very similar to the type they had at home: short-tailed and primitive. Sheep bones from both pre- and post- Viking sites have the same characteristics, "suggesting either that the Vikings brought no sheep with them, or that their sheep were very similar to those already here," and Kinsman feels that there would have been very little point in bringing sheep to Cumbria because there were already plenty in the area of a similar sort.[5]

There is, however, some consolation for the advocates of the Scandinavian connection. There is some evidence that Herdwick sheep share some genetic endowment (shown through gene frequencies for high blood potassium) not only with sheep which are hornless in the female and which are white or tan-faced (such as the Shetland and the Welsh Mountain), but also with sheep found in Scandinavia, such as the Old Norwegian and to a lesser extent the Spaelsau. Herdwicks, however, also share some genetic endowment with various blackfaced types. Herdwicks have a high frequency (0.80) for haemoglobin A, which is identical to that for the Scotch Blackface and slightly lower than those for the Swaledale and the Dalesbred at 0.82 and 0.86 respectively. Ryder interprets this high value for the Herdwick as indicating possible recent introduction of black-faced blood and this is perhaps borne out by the fact that whereas the Herdwick was in 1800 a short-wooled breed, it now has a long and hairy fleece. He adds, "The Herdwick differs from the rest of the group in having heterotype hairs which I originally thought might have come in recently with a long tail from the blackface horn, but it seems that some Scandinavian sheep have heterotypes."[6] For Ryder an alternative explanation of the fact that the black-faced breeds and the Herdwick have a similar level of haemoglobin A, might be due to "selection in a similar environment." The picture then, is a mixed one – on the one hand this, on the other hand that.[7]

An Evolving Breed? The eighteenth and early nineteenth centuries

There is much good sense in this, as knowledgeable local writers have stated over the years. Various people who knew the reality of sheep farming in the fells realised that breeds do not arise naturally and that influences on them are very varied. The extremely knowledgeable Windermere vet, F. W. Garnett, writing in 1912, provided much evidence of this in his book on the agriculture of Westmorland, detailing how in his lifetime distinct breeds emerged in Cumbria.[8] R. H. Lamb, the secretary of the Herdwick Sheep Breeders'

Association in the 1930s, after detailed research, concluded of Herdwicks that, "I am strongly inclined to think that the breed is indigenous and simply the result of a gradual process of evolution." The reality is probably not so much to do with natural evolution but rather to the preferences of a community deciding, over about a century and a half, what they wanted the breed to be like. Behind that process of evolution was a community deciding what constituted the Herdwick breed rather than a Herdwick type of sheep.

The earliest serious description of Herdwick sheep by expert agricultural writers was that provided by J. Bailey and G. Culley in their report produced in the 1790s for the Board of Agriculture and Internal Improvement under the title *A General View of the Agriculture of the County of Cumberland*. There were two kinds of sheep in Cumberland and they were "probably something related." One of the breeds was, "peculiar to that high, exposed, rocky, mountainous district, at the head of the Duddon and Esk rivers" – around the Hardknot and Wrynose and Scafell areas. The mountain sheep kept in these areas were "commonly called Herdwicks." They continued, "the ewes and wethers are all polled or hornless, and also a great many of the tups; their faces and legs speckled; but a great proportion of white, with a few black

Drawing of a Herdwick ram from Bailey and Culley's 1790s report on the agriculture of Cumberland.[10]

spots on those parts, are accounted marks of the purest breed, as are also the hornless tups; they have fine, small, clean legs." Bailey and Culley felt that, from the evidence of kemps or hairs being in the wool, and because some of the rams had spiral horns, there had been some crossing with the black-faced heath sheep. These Herdwicks, however, were "lively little animals, well adapted to seek their food amongst these rocky mountains." The ewes were kept as long as they would breed lambs and were often from ten to fifteen years of age before they were sold. Ewes and wethers (which were sold at four and a half years) were "sold from these mountains, and killed without being put on any better pasture. A wether carcase they saw at Ravenglass weighed 11lb a quarter and had 10 or 12lb of tallow. Ewes weighed from 6 to 8lb a quarter; and a fleece weighed about 2lb."[9]

Their wool was finer than that of the black-faced heath type. They were not fed hay in winter – and indeed to this day fell-going Herdwicks can be reluctant to 'feed'. Bailey and Culley took issue, however, with the often-repeated legend started by James Clarke in his *A Survey of the Lakes* published in the 1780s, that Herdwicks acted, "contrary to all other sheep I have met with" to face the coming storm. Bailey and Culley, however, said "They do not face the coming storm, as reported, but, like other sheep, turn their back on it."[11]

Bailey and Culley, who were amongst the leading experts on farm livestock in the late eighteenth century, also bemoaned the fact that on the common land systems of the fells (which of course prevailed in Cumbria at that time even more than they do today) that even an enlightened breeder would have found it difficult to improve the quality of his sheep, "while his ewes mix promiscuously with his neighbour's flocks." They continued, "If he had the best tup in the Kingdom, can he be sure that one of his ewes would be tupped by him, while there are probably not less than a score of his neighbour's to contest the females with him."[12]

As Garnett also pointed out it took a long time before there was regulation preventing the presence of unsatisfactory entire male animals (inferior rams) on the largely open upland grazing commons. He felt that the unregulated condition of the commons and fells made it, "impossible for any but a heterogeneous type of sheep to be kept." It was not until 1907 that an Act of Parliament was passed prohibiting the pasturing of entire sheep on commons and other unenclosed land.[13]

Reviewing this situation over 100 years later, the great agriculture writer R. E. Prothero (Lord Ernle) set out the problem in the following memorable way: "Stock-breeding as applied to both cattle and sheep, was the haphazard union of nobody's son with everybody's daughter."[14]

Herdwick sheep, like all sheep breeds, are a product of the needs and preferences of the community of people who breed them. Sheep breeds are also the result of that community's abilities to manage the breeding programme for their livestock in the way that it wanted. It is clearly the case that for much of the nineteenth century Herdwick sheep were not very much like the animals we see today, and that there was a great deal of crossing between different types found in the area. The early nineteenth century's foremost expert on sheep, William Youatt, writing in 1837, stated that Herdwicks ("the most valuable sheep on the Cumbrian mountains") were: "small, active and polled, and their faces and legs speckled, having a great proportion of white, with a few black spots strewed upon it."

Youatt's enquiries showed that the lambs were born in May and that the meat possessed "superior flavour." The principal value of the sheep was their hardiness, and he also noted that they were, "once much sought after by the breeders of the mountain sheep." Like Bailey and Culley a generation earlier he suggested that there was much evidence of cross breeding of Herdwicks with the native mountain sheep which produced a sheep that was, "variously coloured about the head and legs, some being white and some being speckled and not a few being perfectly black." These sheep were, "horned, high shouldered, narrow backed, flat sided with coarse and rather long hair."[15] Bailey and Culley seem to have felt that there was a good deal of indiscriminate use of a type that seems to have been emerging from the black-faced type common in Westmorland. These had, they felt, "nothing to recommend them (in our opinion) but *size and coarseness*." On a wet day they looked more like goats than sheep.

This was clearly an era of substantial, often accidental, cross-breeding: but there is also clear evidence that there are pure Herdwicks as well. There was good Herdwick stock to be had – most notably in the Eskdale area. Bailey and Culley felt that people should either pay more attention to trying to improve their breed or, "by hiring or buying some of Mr. Tyson's BEST FORMED and FINEST WOOLLED Herdwick tups." They were referring to the stock of William Tyson, whose family had been tenants on a farm on the Muncaster Estate for four hundred years. This was, they reported, the principal flock of the breed. On account of his flock's *"hardiness of constitution"* Tyson had earned a strong reputation as a great breeder of tups.[16]

This issue of "hardiness of constitution" is a vital point. A central feature of the evolution of the Herdwick sheep was its local adaptation to the climate and terrain. Sheep going on the western and central fells need to be active in relation to the terrain and able to cope with very high rainfall. Practical farmers will have knowingly selected from the animals which thrived best in the situation in which they found themselves. S. D. Stanley Dodgson, in an article called "The Herdwick Sheep, their origin and characteristics," which was written in the early 1920s and reprinted in the flockbooks of that time, asserted that, "it remains an undisputed fact that no breed of sheep has yet been found to stand so well the rigours of the climate on the Cumbrian range of mountains."

The Fell Dales Association for the Improvement of Herdwick Sheep

Breeds are made by people. The British Isles are particularly rich in sheep breeds, there being at least 40 distinct breeds, often very locally adapted and formed from the preferences of groups of farmers. The Herdwick breed of sheep, more or less as we know it today, was created by a community of fell farmers in the Lake District over a period of about 100 years. Selecting animals thought to have superior characteristics and what was deemed to be the right appearance began to be a particular concern of nineteenth century Lake District sheep farmers. In the days before improved communications, especially a railway network, and the establishment of auction marts there were numerous local sales and fairs at which ewes, wethers and lambs were brought from the high lying areas to be sold into the lowlands. In the west of Herdwick country, for instance, there were seasonal sales at Boonwood near Gosforth and at Lamplugh – with every other area having its fairs and sales through which the product of the fells was sold to farmers on lower ground.[17]

Throughout the nineteenth century, for instance, Ennerdale Bridge had an annual sheep fair held on the second Tuesday in September. In addition, Gillerthwaite, the great dalehead farm of Ennerdale, annually advertised its sale of, "Draft sheep from the Coves" which it held by the side of Ennerdale Water at Bowness.[18] These were the great income earning sales of the year, "the harvest of the hills" – but, given as they say that "tups are half the flock," there had also to be specialist sales for fell tups. To give just one example, Sir Daniel Fleming of Rydal Hall, recorded in his accounts for 1695 that he paid 16 shillings for a tup at Ambleside Fair.[19]

In 1844 the West Cumberland Fell Dales Sheep Association was established largely for the purpose of enabling breeders to get access to fell tups either to be bought or to be hired. The show and sale which covered the 'Above Derwent' area (that is the area between the Derwent and the Duddon rivers) rotated in those early years between Ennerdale, Loweswater and Nether Wasdale, and usually attracted about 100 exhibitors.[20] In 1848, when the West Cumberland Fell Dales show was taking its turn to be at Ennerdale Bridge, Thomas Rowlandson wrote a vigorous account of it in the *Journal of the Royal Agricultural Society of England* (Vol.X), in an article entitled 'On the Breeding of Sheep best adapted to different localities': "The full attendance of breeders and agriculturalists generally, caused the trade in tups to rule with unusual activity, and it is computed that not less than two hundred specimens of the genuine breed changed hands on that occasion."

But what did these Herdwicks look like? Rowlandson gives a description of the breed and indicates its evolving nature:

> "The Herdwicks of the present day are characterised by being polled, and have brownish or speckled black and white or mottled faces; some few have black faces, and some also have horns, but neither of these are considered genuine; they are also known for the circumstance that, as they get older, they assume a white or grey appearance about the nose and legs (in the shepherds' phrase they grow raggy). The ewes should always be polled, on a few wethers and rams small smooth horns make their appearance – a proof of admixture of blood…"

There were 94 competitors "exclusively devoted to one description of sheep, all of whom dwell within a very limited extent of country." The best tup was shown by Robert Briggs of Wasdale Head. Other prizes were won by Edward Nelson of Loweswater, John Bowman of Mireside, Ennerdale, Joseph Tyson of Tows, Eskdale, Thomas Pearson of How Hall, Ennerdale and J. Tyson of Gillerthwaite, Ennerdale. Significantly, showing the variation in the breed at the time, there were classes for five white fleeced ewes, (with Pearson, Nelson and John Jackson of Swinside End, Ennerdale, taking the prizes). Also there was a prize offered by the Keswick Manufacturers for the "best white-fleeced sheep of the genuine Herdwick bred and depastured in Allerdale, above Derwent."

In 1869, even after he had emigrated to the United States, Rowlandson was still writing about Herdwick sheep, and suggested that at about the turn of the century some individuals had begun to think about trying, "to breed a pure race, their character having become much intermixed by crossing with the indigenous 'black-faces'."[21]

Other contemporary accounts bear this description out. For instance, William Dickinson, the prize winning west Cumbrian agricultural improver and writer, in 1851 states:

> "The Herdwick breed possesses more of the characters of an original race than any other in the county. It stands lowest on the scale of excellence, and shows no marks of kindred with any other race... The majority are

John Wilson of Wasdale Head with champion Herdwick ram 'Nero', 1870.

without horns, and their legs and faces are grey or mottled. Where great care is exercised in breeding, the nose is of a lighter grey, and is then termed "raggy" or "rimy", from its resemblance to hoar-frost."

Dickinson also described as living on Skiddaw, Saddleback, the Caldbeck fells and on the Helvellyn range, "another breed resembling the Herdwicks, but stronger in bone and heavier in carcase, which the owners claim to be a superior and distinct breed from the "little Herdwicks", as they derisively call them... These have not acquired any local appellation."[22]

Another description occurs in a history of Cumberland and Westmorland published in 1860 where it is mentioned that a "peculiar breed of sheep, called Herdwicks, from their being farmed out to herds at yearly sum, is met with on the mountains, at the head of the Duddon and Esk rivers. The ewes and wethers, and many of the tups are polled; their faces and legs speckled, and the wool short. They are hardy little animals. The tups are in great request, to improve the hardiness of other flocks."[23]

Certainly the engraving of two Herdwick tups accompanying an article in *The Field* in 1873 would confirm the physical description. These two tups – which came from Christopher Wilson's Kentmere Hall flock, and one of which had come second to a Lonk tup at the Royal Show in Hull that year – look for all the world like fairly ordinary black-face heath type sheep with black and white muzzles.[24] Similarly, Canon H. D. Rawnsley, may well have been observing the same phenomenon when he wrote about the father of Joseph Hawell of Lonscale having built up a flock of 500 "black-faced Herdwicks" on Frozen Fell and Wylie Ghyll and 300 in Skiddaw Forest.[25] Again, in his piece 'A Crack about Herdwick Sheep' published in 1911, he wrote of, "the delicate, lithe, little sheep with their shy black faces and their dainty feet, that give life to the mountain side"[26] The only safe description of a Herdwick at about this time was probably that it was a fell-going or heaf-going sheep.

This fell-going or heaf-going ability was a key determinant. A handbill for the annual meeting and tup fair of the West Cumberland Fell Dales Association, when it was held at Nether Wasdale in October 1850, included amongst the rules the following very revealing one:

"The Judges are requested to reject any Sheep which they may consider to have been unfairly flushed for the sake of exhibition, the main purpose of the show being to exhibit the most useful Sheep for a Heath-going stock."

After an existence of about twenty years, a new institution was created: the Fell Dales Association for the Improvement of Herdwick Sheep. This was established in 1864 when its first show was held at the Woolpack Inn in Eskdale. It was without doubt a key institution in creating the Herdwick breed, not least with its emphasis on insisting that the Herdwick was fit for purpose as a hardy sheep capable of going to and prospering at the fell. 'Eshd'l Show' (Eskdale Show) certainly became the largest gathering of Herdwick tups for sale but largely for hire, though the 'Below Derwent' autumn tup fair continued to be held at various locations in Keswick.[27]

Rules to enforce the requirement of fitness for mountain life were kept going after the Fell Dales Association was formed. In the 1876 schedule of prizes, for instance, there was a special prize for twenty ewes and twenty lambs "all direct from the Fells in the local district" with the stipulation that "no field fed sheep, nor any which goes on enclosed allotments of common land will be allowed to compete for the prizes."[28] The importance of ability to go the fell was still reflected at the 52nd Show in 1916 when the catalogue was issued in the name of the 'Fell Dales Association for the Improvement of the Herdwick and Fell-Going Sheep.' Herdwicks, in short, were fell-going sheep – sheep that could live on the mountains – or as the Reverend Ellwood of Torver put it in 1899, "Our Mountain sheep... they are called Herdwicks."[29] Ability to thrive at the fell was the great priority rather than simple improvement.

But, increasingly, leading breeders began to lay down breed points. Classes for Herdwick sheep (which were often described as Mountain sheep in the schedules) began to become common at agricultural shows. F. W. Garnett researched this in detail for Westmorland and east Cumberland, illustrating also the easterly extent of Herdwick sheep (though often alongside black-face heath types both on the fells and in the show pens). There were classes for both Herdwick sheep and black-faced sheep at the shows at Penrith from 1833, Appleby and Kirkby Stephen from 1841, High

Prize winning Herdwicks, Royal Show, Newcastle, 1864, polled but speckled.

Borrowbridge (near Tebay) from 1848, Staveley from 1851, Orton from 1860, Shap from 1861 and Kendal 1865.

Two other organisations also played a significant role in improving the standard of Herdwick sheep. The Troutbeck Herdwick and Other Sheep Association was formed in 1868 but ended in 1875 after a bad local outbreak of foot and mouth disease. At the annual show there were fourteen classes for Herdwicks and four for white-faced tups for getting cross-bred lambs out of older or 'crock' ewes. William Dickinson, the west Cumberland farmer and agricultural writer, recorded in 1850 that this practice generated, "lambs for sale worth 50 per cent more than lambs of the true mountain stock."[30] The Leicester in particular was for many years used as the sire of a very successful cross-bred lamb.[31] There were also classes for Leicester tups at the Fell Dales Association shows, for instance in 1875. Associated with the old Bampton Tup Fair, there was also the Bampton Association for the Improvement of Mountain Sheep which started in 1876 and lasted until 1898. There were six classes for Herdwicks, and ten for black-faced sheep.[32]

In the east, the breeds which were emerging from the black-face heath type (the Swaledale, the Blackface and especially the Rough Fell in the area round Kendal) were kept alongside Herdwicks. In these areas enclosure of the rather easier fells was more feasible and made flock and breed management more possible. In an account of the Kentmere valley in the 1890s, there is a description of the heaf-going flock (1500-2000 strong) at Kentmere Hall where, "the breed of sheep which find food upon the mountains are either the Herdwicks or small black-faced, which are said to be peculiar to the district."

We can almost certainly see here in this black-faced group the beginnings of the emergence of the Rough Fell breed. This source also gives a plausible early account of the sheep 'peculiar' to the fells around. "By selection, the special characteristics which fit it for its elevated home are reproduced again and again, until now they became fixed."[33] The enclosure of the lower slopes of the fells began to make it possible for the two breeds of the area to begin to diverge rather than to mix and re-mix continually: a sharp contrast to the situation described in Westmorland a hundred years earlier by A. Pringle in his *General View of the Agriculture of the County of Westmoreland,* when he wrote that the breed of sheep on the mountains and commons of the county were, "either native or a cross with Scotch rams... horned, dark or grey faced, thick pelted, with coarse, strong, hairy wool."[34]

'Essays on Herdwicks': Defining the Breed Type

The publication of a new *Shepherd's Guide* compiled by Daniel Gate in 1879 for the fells of Cumberland, Westmorland and Lancashire, offered the opportunity for renowned breeders in some 'Essays on Herdwicks' to lay down breed points and to discuss the merits of the Herdwick breed. William Dickinson, whose memory went back to the first decade of the nineteenth century, asserted that there had been a serious reduction in the extent of Herdwick sheep-keeping; that it was now nearly confined to the west Cumberland range of mountains between the Derwent and the Duddon rivers. Sheep outside this area were larger-framed as a result, however, of better winter treatment and better summer pasture than was available on the rocky pastures of the west: but these were not "pure and hardy heaf-going Herdwicks." He felt there was no satisfactory evidence about the origins of the breed – though he did tell and effectively dismiss the Spanish shipwreck story, arguing that the hardiness of the Herdwick was much more likely to have originated from northern climes such as Sweden or Denmark. He also recorded that the nearest thing he had ever seen to a Herdwick elsewhere was a sheep rather like a Shetland type he had seen in the Morne mountains on the west coast of Galloway about fifty years earlier.

Another contributor, William Abbott of Coniston, wrote an extremely enthusiastic account of the merits of the Herdwick. He noted that although there had been some criticism of the breed in the farming press in favour of improved breeds, he was having none of it – arguing that there was, "not a breed anywhere in the wide world capable of taking the place of the Herdwick." A good coat was extremely important and he also felt that, "there should be no specks upon the faces or legs; these are not belonging to a purebred Herdwick."

Edward Nelson of Gatesgarth and John Wilson of Keskadale contributed the final essay. For them, coat was at the top of the list – no doubt reflecting the economic value of the wool crop – and they said that the Herdwick was previously, "an ill-developed, coarse-wooled animal" but was now "well-formed" and "covered with good wool." They were looking for a barrel-shaped body, with legs set well to the outside, of strong bone, and with a broad head. There clearly were problems still with horns because they felt obliged to mention that these were "not desirable in the female."

The Edward Nelson who was co-author of those opinions was more usually referred to as Ned Nelson (senior), someone whose influence on the development of the breed was monumental. Ned Nelson was a son of George and Ann Nelson. The Nelsons were originally from Caldbeck but they had also farmed briefly at Chapel House, Borrowdale, in the 1820s. Not having done very well at Chapel, George Nelson went to Patterdale to manage Patterdale Hall for 'Squire' Marshall. When Ned Nelson was twelve he was hired at Keskadale in Newlands and later moved to Miss Skelton's employment at Foul Syke, Loweswater, where he worked for eight years.

He then married Eleanor Banks who was daughter of John Banks who kept the Kirkstile, Loweswater, which was both a farm and a public house. Over a fourteen year period he built up a substantial flock at the Kirkstile. He took 200 Kirkstile sheep with him when took on the tenancy of Gatesgarth, Buttermere, in 1850 at the age of 35. He there took over a stock of 1,447 sheep: ewes, hoggs and wethers, and dedicated himself to improving his stock and the farm on which they were kept. He employed Irish labourers to drain Warnscale bottom, and at his behest, they also straightened Crookabeck. He cleared over 600 acres of fellside intake of bracken. He planted trees for shelter and dug gravel from the lake shore to build hard roads. He built a large barn measuring 11 yards by 33 yards, and a great wool shed. There were only 160 acres of inbye land, but huge sheep heafs eight miles long and four miles broad on both sides of the valley, making a total of about 3,000 acres. There were (and still are) four separate stocks of sheep: Gatesgarth Side, Birkness, Scale Force and Fleetwith, comprising a combined flock of 2,500 ewes plus followers.

By the mid 1870s there were reportedly 150 Herdwick tups available annually for hire at Gatesgarth. Nelson let this number of tups at a fee of about 2 to 5 guineas each and sold one tup for 30 guineas at a time when 12 guineas was a tremendous price. Another of his most celebrated tups, 'Boggle,' was let for the very large sum of £1 per day for seven days and a further ten days for £7. Andrew Humphries in assessing this evidence, has calculated that this means that one cycle with the ewes was costing perhaps the then large amount of 10 shillings for each lamb produced. He commented that breeders such as Nelson, "clearly influenced large numbers of blood lines and through the retention of ownership of most of their rams retained control and flexibility in the emphasis on various key strains."[35] It was in this way that the larger farms, especially those at the daleheads, exerted their strong influence on the breed – not least due to the fact that they often held what were called 'Tip Sundays' on their farms when other breeders were

Gatesgarth, early 1930s - Allan Nelson, right, and Syd Hardisty, centre.

invited (before the Fell Dales show took place) to inspect tups and to "speak for them" for the winter.[36]

Ned Nelson senior, was very committed to his breeding programme and exhibited widely at agricultural shows. After years of effort and persistence a Herdwick of his won the sheep championship at the Royal Show when it was held in Newcastle in 1864.[37] His son, also Edward (Ned) who was born in 1847, took over the farm in 1887 and had a similar enthusiasm. In 1920 he was described as having "even excelled" his father with the great strength in depth of his Herdwick flock. It was said that, "it is probably literally true that in every known Herdwick flock there is a strain of Gatesgarth blood" and that his 500 best breeding ewes vied with each other in showing, "all the characteristics of the true Herdwick."[38] In the end, however, he could scarcely make it pay. He retired from the farm in March 1934 and died in September 1934 worth only £17 after his debts were settled.[39]

Although Herdwicks were inevitably talked up by their enthusiasts, not everybody in the agricultural community was so enamoured. A writer in *The Field*, for instance, having seen Herdwicks at the Royal Show in Carlisle in 1880, complained that, "They look like the last remnant of, we won't say barbarism, but of very ancient and primitive sheep breeding." At about the same time Finley Dunn in the *Journal of the Royal Agricultural Society of England* wrote that, "the black spotted goat-like Herdwicks are still susceptible of considerable improvement."[40]

The *Shepherd's Guides* of the mid nineteenth century reveal that the huge majority of the Lake District sheep stocks were of a Herdwick type, that is to say that they had no horns in the female. It is only in the flocks on the eastern edge of the Lake District that sheep are mentioned as being marked by horn burns and not, therefore, of a Herdwick type which is a polled sheep. The 1849 *Shepherd's Guide* compiled by William Hodgson of Corney, uses illustrations of hornless Herdwick type sheep, to show the marks. The *Guide* reveals that 23 of the 27 Crosby Ravensworth stocks had horns, as did seven of the fifteen Rosgill and Shap stocks, and all but one of the 25 Birbeck fells stocks, along with eleven of the eighteen Swindale stocks. Longsleddale had eleven stocks of horned sheep and nine stocks of sheep without horns. There were other valleys on this eastern edge where there were a mixture of the horned and polled types, such as in the Haweswater area (Measand and Mardale) where twelve stocks were of Herdwick sheep and only three (two stocks at Mardale Waters and one at Chapel Hill) were of a horned type.

The 1879 *Shepherd's Guide* itself (with an engraving of what is clearly a Herdwick tup to illustrate the far side of the sheep and a Herdwick ewe to illustrate the near side of a sheep and to show the earmarks) gives us some tantalising clues of how the Herdwicks and the black-faced types were kept cheek by jowl with each other especially in the old north Westmorland. In the long entry on the Swindale, Wetsleddale, Rosgill, Shap, and Birbeck Fells area, which contained 46 stocks of sheep, only nine of them do not have horn burns – indicating that the ewes must have had horns and therefore were black-faced type sheep. Similarly only four of the seventeen stocks in Longsleddale do not have horn burns. In the chapter on Bampton, Measand and Mardale, there are (among others) two entries for stocks of sheep managed by William Dargue. The entries are revealing. The Thornthwaite Hall stock was, "black faced, stroke down far shoulder, pop on near hook, T on near horn, the year of our Lord on far horn." His other entry read, "Naddle sheep, cropped both ears, burned across face. Herdwicks D on face."[41]

The ability with which farmers could differentiate one breed from the other clearly increased in part as a result of the enclosure of substantial areas of common land. Not only did enclosure help to reduce the widespread phenomenon of the over-stocking of commons (which in turn led to a great deal of 'dogging' of sheep)[42] it also enabled breeding programmes to be better regulated, due to farmers having more land under their sole and direct control – a point that was made earlier. It was precisely in these decades from the 1840s up to the 1870s, when Cumbrian breeds were clearly emerging, that enclosure reached a high point in Cumberland and Westmorland. In fact, the two counties accounted for thirteen per cent of all land enclosed between 1820 and 1870.[43] The occasion for this late flurry of activity was no doubt the continuing pressure on grazing resources brought about by high demand for agricultural products in the country generally.

The enclosure phenomenon did not, however, involve anything like the whole area of the fells. Some examples from part of the west Cumberland area of Herdwick country will serve to make the point. In Lorton, for instance, the fells were enclosed in the 1830s but a huge area of the neighbouring fells between Brackenthwaite and Braithwaite remained open. When the Lamplugh fells were enclosed in the 1840s, the terrain was such that it was possible to turn it all into private property. Loweswater parish followed in the 1860s and again the terrain was such that all the high fell area could be enclosed using the new products of metal posts and wire for boundary purposes. The enclosure of the Ennerdale fells followed in the 1870s, but this was almost certainly too ambitious as it became impossible to sustain the fences on Ennerdale's large areas of crag and shifting scree. The neighbouring townships of Kinniside and Nether Wasdale, however, remained unenclosed – even against each other. In short, Herdwick country retained much of its open fell character.

Where enclosure was successful and sustainable it clearly made a difference. The first president of the Herdwick Sheep Breeders' Association, S. D. Stanley-Dodgson (who was William Dickinson's grandson) developed a high quality Herdwick flock on the enclosed Harrot fell near Embleton, bred up from sheep from a variety of farms. He also, in the first Flock Book in 1920, made it clear that as a result of enclosure, "the breed of Herdwicks has been much improved by careful mating." Writing of what for him were the bad old days of common land, he stated that, "Negligent breeders turned very moderate rams loose, with baneful effect on the breed of a district." In addition it was no longer necessary to keep wethers to defend the heaf, a point also made by another advocate of fell enclosure, the Kendal land agent Crayston Webster, whilst regretting the loss of four year old wether mutton that would accompany it.[44]

It is clear that many Herdwicks at this time still did not look completely like Herdwicks of the present day. F. W. Garnett wrote in 1912 that, "Pure Herdwick lambs are beautifully white, with black legs and faces and white ears – though often there are black spots on their bodies, which afterwards die out."[45] Twenty or so years later R. H. Lamb confirmed this, "Herdwick lambs when born are nearly always a good black and white, the head and legs being black, the rest of the body being white or blue spotted, but as they grow older the dark colour gradually changes to a light grey." The "sooty faced, pie-bald" lamb at a year old has a brown fleece and a grey face. When mature the wool is grey and the face 'rimy' or hoar-frosted in appearance.[46] By contrast, the majority of present-day breeders would prefer a lamb to be largely black on the body and head, perhaps with some flecks of white in the body and on the ears.[47]

Canon Rawnsley and Herdwick Sheep

An interesting attempt to form a Herdwick Sheep Association was made in 1899. S. D. Stanley Dodgson of Tarnbank, Cockermouth, and Noel Rawnsley of Crosthwaite called a meeting of Herdwick breeders which took place in Keswick Cricket Pavilion on 6 September. The meeting was chaired by Noel's father Canon H. D. Rawnsley, one of the triumvirate who had formed the National Trust for Places of Historic Interest or Natural Beauty – the National Trust – in 1895. Canon Rawnsley, the vicar of Crosthwaite (Keswick), was already the driving force behind the Lake District Defence Society, an organisation which campaigned against what he perceived to be inappropriate development – whether of railways, road improvements, quarries and the like. Rawnsley was famously 'into everything' and campaigned for public access to mountains, agricultural education, the preservation of customs, the relief of tuberculosis, the development of secondary education – and much more.[48]

He was elected to chair the meeting, having been proposed by D. N. Pape and seconded by Mr. Rothery of Whinfell. The Canon said that the purpose of the meeting was to get an expression of feeling as to whether it was worth pursuing the formation of a Herdwick Sheep Association. Rawnsley went on to suggest that it was: mentioning the price of wool, the general growth of sheep numbers in the country, the potential of Herdwick mutton and also that, on the first occasion that Herdwick sheep had a class of their own at the Royal Show, the people organising the show so lacked knowledge about the breed that they, "gave the judge no specific instructions as to points."

Founding fathers of the Herdwick Sheep Breeders' Association at the Fell Dales show, 1916. Ned Nelson with beard and trilby, centre front row.

Noel Rawnsley followed, speaking on behalf of himself and Stanley Dodgson, and set out a proposal to form a Herdwick Sheep Association on similar lines to associations that had been formed in recent years for lowland sheep of various breeds. They wanted the objects of the Association to be: 1) to improve and maintain as a distinct breed the pure-bred Herdwick; 2) to spread information about the breed and make its excellent mutton and hardy characteristics more widely known; 3) to establish a flock book; 4) to obtain special classes for exhibition of the breed in other than local shows; 5) to draw up a specification of the points desired in a typical Herdwick; 6) to issue periodically a 'smit and ear-mark register.' After a couple of speeches and a small amount of debate – with one speaker (Mr Abbott of Beckstones) stating that he felt Herdwicks were on the decline and another (John Wilson) saying there was nothing to lose – it was agreed to form a Herdwick Sheep Association.[49]

This Herdwick Sheep Association subsequently held ram fairs in the spring and at the back end of the year, and it probably formed the basis of the Keswick Ram Fair which still meets today in mid-May, but it did not seek to establish a flock book. The organisation which took that crucial step, the Herdwick Sheep Breeders' Association, did not result from this initiative. The Association was not formed until 1916.[50] It seems likely perhaps that the Rawnsleys' attempt to form an all-embracing Association did not have enough grassroots support and that the legendary independence of mind of the Herdwick farmers meant that they wanted nothing to do with yet another enthusiasm of the energetic founder of the National Trust. They clearly would not work to the Canon's complete agenda and timetable and in the event waited until they were ready: a typical act, perhaps, of self-determination.

There is, however, little doubt that Rawnsley was a close observer of the world of Herdwick sheep keeping, and he expertly described that world in his, 'A Crack about Herdwick Sheep' written in 1911. He covered with a good degree of accuracy and completeness such things as the heafing and homing propensities of the sheep; the fact that Herdwick mutton was, "the sweetest of its kind in Great Britain"; the pattern of sheep management throughout the year; about lug and smit marks and the *Shepherd's Guides*; problems with maggot fly; communal activity at clipping time; the importance of sheepdogs; salving and dipping, and problems with snow. Above all he seems to have lionised the shepherds themselves: they were a "fine race these Viking shepherds... We still have among us the Michaels that Wordsworth knew and described."

It is very plain to see in any of Rawnsley's writings about Herdwick sheep that he was full of admiration for the people who kept them. He described the 'cultural landscape' and appreciated it. Traditional farming activity and landscape preservation were, as far as he was concerned, the same side of coin. He concluded his 'Crack' with a section on sheep scoring numerals which, he observed regretfully, had fallen into disuse –and surely revealing some of the apocryphal nature of this quaint system, he indicated that, "I have never been able to find that these numerals have been used by shepherds of our own time. The oldest men I have spoken with could only say that their fathers told them that their fathers always counted that way." He went on to say, however, "we who dwell in the land of the shepherd must view with regret any passing into oblivion of shepherd customs or shepherd speech."[51]

Forming the Herdwick Sheep Breeders' Association

It is hardly the case in the years before the First World War that the breed was in any difficulty – in fact, it was probably at a high point. One reliable estimate from this time suggests that there were about half a million Herdwick and Herdwick-cross sheep, making up about 40% of the sheep population of Cumberland, Westmorland and Lancashire north of the sands.[52] The Fell Dales Association, which had a crucial role in the supply and improvement of Herdwick tups, was flourishing. At the 1910 Fell Dales show, for instance, held in Eskdale on 30 September, there were 21 classes. Included amongst them was a class for the best ten tups for sale or hire which attracted 57 entries. Garnett analysed this very large entry, finding, incidentally, that of the 570 tups, twenty or about four and half per cent were without horns, though, "many others had horns of diminutive size."[53]

The 1916 catalogue lists an amazing 31 entries in this same class, including ones from J. Rothery and R. Wilson both from Wasdale Head; J. Roper of Bowderdale, Wasdale; J. Birkett of How Hall, Ennerdale; Wm Birkett of

Mr and Mrs Harryman of Keskadale and dog. Mr J Harryman chaired the inaugural meeting of the HSBA.

Gillerthwaite, Ennerdale; Isaac Thompson of Wythburn; J. Plaskett of Borrowdale; Jeremiah Richardson of Pike Side, Ulpha; J. Harrison of Brotherilkeld, Eskdale; Simpson Richardson of Taw House, Eskdale; J. Williamson of Routen, Ennerdale; J. Richardson of Swinside, Ennerdale; J. N. Gregg of Troutbeck, Windermere; E. Nelson of Gatesgarth, Buttermere and W. N. Park of Crag, Ennerdale.

Many of these well-known names in the Herdwick world were founder members of the Herdwick Sheep Breeders' Association. A large group of Herdwick breeders (who were photographed as a group on the Fell Dales showfield) had decided to call a meeting with a view to starting an Association – presumably motivated by a determination to improve the standard of the breed.

An advertisement inviting people to the meeting was placed in the *West Cumberland Times* by William Tyson of Watendlath. A few weeks after the 1916 Fell Dales Show the inaugural meeting of the Association was held. The meeting took place on Saturday, 4 November 1916, at the Royal Oak Hotel, Keswick with J. Harryman of Keskadale in the chair.

S. D. Stanley-Dodgson, a farmer and landowner from the Cockermouth area, who had been involved in the Rawnsleys' attempt to set up an association, was elected president. William Wilson of Watendlath ('Herdwick Billy' as he became known) became secretary.[54] Bob Devon of Bram Crag, St Johns in the Vale, became treasurer and Joseph Broatch was appointed as

Origins and Development of the Herdwick Breed

the honorary legal advisor. No doubt partly because of it being war-time, the first flock book was not produced until 1920, with the opportunity being taken to give a potted history of all the Association's flocks. Nearly 1,200 rams were registered on a self-inspection basis by the breeders and the original 130 farms they came from covered a wide area of the Lake District fells.

The flocks ranged in size from the great dalehead flocks from farms such as that of John Edmondson at Seathwaite in Borrowdale (Flock No.1, with 500 ewes);[55] William Wilson's Watendlath flock (No.5 with 800 ewes); Joseph Harrison's flock (no. 13 with 500 ewes) at Black Hall, Ulpha; Isaac Thompson's one thousand strong West Head flock (no.38) which had been replenished with 100 good ewes from Gillerthwaite about 30 years earlier; William Birkett's Gillerthwaite flock (No.42) in Ennerdale ("one of the oldest established ram breeding stocks" with 606 ewes put to the ram); Joseph Richardson's Swinside, Ennerdale, flock (No.46 with its 500 ewes in three separate stocks: the Allotment stock the foundation of which came from Gillerthwaite in 1911 when he took the farm, the Swinside stock sheep which went on Lankrigg and the Yargill stock also on Lankrigg, Kinniside's main fell); Edward Nelson's Gatesgarth flock (No.52, 500 ewes) in Buttermere ("one of the oldest, largest and best-known throughout the Herdwick country") and Richard Wilson's Middle Row flock (No.60, 600 ewes) at Wasdale Head ("a noted Ram Breeding Flock of old-standing"). Tyson Hartley of Turner Hall, Seathwaite had 250 ewes in Flock 14 with the

1st PRIZE HERDWICK SHEARLING EWES,
Royal Show at Newcastle, 1908,
the property of The Right Hon. The Earl of Lonsdale.

Turner Hall prefix and the same number in Flock Number 23 with the Mosshouse prefix.

There were 408 ewes kept in Ennerdale Dale by the Earl of Lonsdale – who had received the dalehead as the manorial share of the Ennerdale fells when they were enclosed in 1870s. Thomas Rawling had a flock of 400 ewes at Lanthwaite Green (No.56); Simpson Richardson had one of the same number at Taw House, Eskdale (No.63); as did Thomas Ridley at Woodhall, Hesket New Market (No.67). Messrs Gregg put 450 to the tup at Town End, Windermere (No.86), and Mrs Leck had a 500 strong flock at Troutbeck (No.87). W. Leck had flock number 88 at Wood Farm, Windermere, with 400 ewes. At Patterdale Hall, the property of the manufacturer William Hibbert Marshall, flock number 124, put 600 ewes in two stocks (one grazing on Thornhow and the other on Moorside) to the tup. It was also stated (describing what was commonly the case) that, "Wethers are kept and are sold at four years old."

John Rothery put 500 ewes to the ram at Wasdale Head Hall, a farm he entered in 1901 and where there were 754 stock sheep, "which have been held continuously with this farm for over 200 years." He also mentioned what are presumably some better quality sheep in a flock he had founded in 1873 with ewes from Allan Pearson's celebrated Lorton flock from the previous generation. Pearson's Bridge End flock along with that of John Sanderson of Beckstones, Thornthwaite, had had a significant influence on several other stocks of sheep. William Abbott of Mockerkin's Blakefell stock had been founded with stock from them and John Bell of Moss Cottage, Loweswater, had strengthened his Galefell stock with sheep from both Pearson and Sanderson, no doubt enabled to manage them after the enclosure of the Loweswater fells in the 1860s.

Simeon Grave of Low Skellgill, Newlands, where the Grave family has farmed since 1347.

This dynamic situation of flocks being created or improved

using infusions of fresh stock from elsewhere was quite marked. This is not an unchanging world – retirements or removals become opportunities for new flocks to be established or for old ones to be taken over by agreement. Lord Leconfield had created his Skiddaw Forest stock by purchasing the flock of pure Herdwicks that grazed that ground from the late Mrs Cockbain of High Row, Threlkeld.

John Cockbain of Causeway Foot had bought from John B. Allison of Low Nest, Keswick, a flock of Herdwick ewes grazing on Helvellyn. Joseph Todhunter at Mirehouse had a stock which was "raised from Ewes bought from Robert Hawell of Lonscale" – another farm that had a large influence on the development of the breed. S. D. Stanley Dodgson, as well as creating a new flock from various good bloodlines to go on the enclosed Harrot Fell, had also taken over Thomas Clemitson's Hobcarton, Whinlatter flock. The flock that belonged to the Oaks at Loughrigg was recorded as having been bought 55 years earlier from the farm at High Close. Others had built up their flocks from small beginnings – for instance, Jonathan Cowx at Town End, Uldale, had started 30 years earlier with "ten of the best Gimmer Lambs from Skiddaw Forest and from Sheep obtained from the late John Harrison, George & Dragon Inn, Uldale."

The great strength of the breed was in the area between the River Derwent and the River Duddon – and Ennerdale in particular was a great stronghold with about one sixth of all tups registered coming from that valley. There were, however, flocks well outside the western and central area. There was,

"OLD PERFECTION,"
bred by and the property of Messrs. Wilson, Watendlath and Wasdale Head.
Winner of numerous prizes.

for example, a flock of 2,000 Herdwick sheep (including 1,000 ewes) at Kentmere Hall and a 500 strong ewe flock at the Oaks, Loughrigg – and numerous flocks on the Skiddaw Range – where the sheep were often thought to be larger.

Throughout the whole Herdwick area there were numerous small flocks of about 250 ewes, like that of William Tyson at Penny Hill, Birker – where there was "the usual flock let with the farm" – or that of Henry Dawson at Folds, Ulpha, which carried "the regular flock of heaf going sheep let with the farm."[56] There were also particularly old flocks which had been managed by the same families since the late eighteenth century – such as that belonging to Thomas Bowes at Fenwick, Thwaites, ("handed down from father to son without a break since the year 1789")[57] and that of the Ponsonby family at White Banks, Kinniside, where the flock had been "in the hands of the present owner and his predecessors since 1773."

It is interesting to note that of the 130 farms in the inaugural flock book, there are only six families still breeding Herdwick sheep on the same farms at the start of the 21st century – although there is a significant number of other families listed in 1920 who are still breeding Herdwick sheep but on different farms. The Herdwick breeding families who are still on the same 'spot' are:

1) Edmondson, Seathwaite, Borrowdale flock No 1;
2) Hartley, Turner Hall, Seathwaite flock No.14;

Derwent Goalpost, champion ram, Eskdale Show. Owned by W. Wilson, Herdwick Croft Farm, Keswick.

3) Rawling, Hollins, Ennerdale flock No.47;
4) Grave, Low Skellgill, Newlands flock No.78;
5) Brownrigg, Millbeck Hall, flock No.96;
6) Troughton, Thwaite Yeat, Broughton in Furness, flock No.116.[58]

In the following year, 1921, a further 900 Herdwick tups were registered from another 67 flocks. The majority of the flocks were from typical small fell farms such as that farmed by Thomas Thwaites of Ghyll Bank, Newlands, with his flock prefix of Catbells. Thwaites put 120 ewes to the tup – with "all ewes bred on the farm, and pure-bred Herdwicks." A less typical stock, showing some persistence of Herdwicks in the eastern part of the Lake District, was that of E. W. Bindloss at Hartrigg in Kentmere where there were 350 ewes "brought from Rosthwaite."[59]

By the time the 1927-28 Flock Book was published the total number of Herdwick flocks in membership of the Association had risen to 245. The newcomers were in some respects a typical sample: Wilkinson Brown of Gillbrow, Newlands, registered "an old established flock of pure Herdwicks" and John Hartley of Greystone House, Thwaites, registered "the regular flock let with the farm." Richard Dalton Weir of Penfold, Dockray, stated that his flock had "been on the holding a long time." There was a much larger claim made, however, for Flock No. 242 that of Robert Wilson of Burnthwaite, Wasdale Head, where 250 ewes were put to the ram. This flock was claimed to be, "One of the oldest pure-bred Flocks in the Country. Many prize-winners, both male and female, can be traced back to this Flock, which has been carefully handled for at least a hundred years." Next door, at Middle Row, Wasdale Head, Joseph Naylor registered a 500 strong flock, No. 241, taken over in March 1928 from R. M. (Dick) Wilson, who had moved to Glencoyne, Patterdale. In the next Flock Book, that of 1929-30, another five were accepted into membership, notable among them being Joseph Harrison of Butterilket, with a 600 strong flock.

The names of some of these families still resound in the world of Herdwick sheep breeding, even though some of them are no longer involved. But there are still Richardsons, Tysons, Birketts, Harrisons, Hartleys and Rawlings and others breeding Herdwick sheep. Other families have changed farms and moved around – often on several occasions. There has also been a succession of new families coming into the Herdwick world over the years indicating that change is a feature even of a world that seems as unchanging as that of Lake District sheep farming. It is also obviously the case that many families, which in the early days of the Association kept Herdwicks, now keep Swaledale sheep at the fell: examples being the Cockbains from the Keswick area, the Pears family of Fellside, Caldbeck and the Cowx family of Uldale to give just a few examples.

Depression and Difficulties

The years immediately after the end of the war were years of economic recovery and reconstruction in Britain. A large group of Herdwick breeders, conscious of the post-war difficulties in continental Europe, magnanimously decided to send sheep to help with the re-establishment of farming in northern France. In March 1920 four tups and 140 ewes from a wide group of breeders were donated to farmers in the French part of the Ardennes to help them re-establish their sheep flocks after the devastation of war.[60]

Shortly afterwards, however, things began to get difficult. At home, 1921 brought the economic depression which did not abate until the period of re-armament in the late 1930s. Cumberland, especially west Cumberland, was in the grip of some of the most difficult economic conditions in the country with male unemployment rates in the mining and steel making west coast towns reaching levels in excess of 50%. In the farming community one phenomenon was the sheer amount of moving from farm to farm which went on. There were dozens of farms to let during these years and people could pick and choose if they had any resources at all and any reputation in order to 'better themselves' on a more promising farm – perhaps indeed one with a better fell, more in-bye and a bigger landlord's stock to take over. Sometimes they moved, I have been told, simply in the hope of a change of luck.[61] For instance, to give some examples which just concern the township of Kinniside, in the early 1930s the Bland family moved to the Nook in Borrowdale from Standing Stones on Kinniside, whilst the Roper family

Wasdale Perivale, bred by R. M. Wilson, Ullswater.

William Bowes, Fenwick, when ten-years-old.

moved from Corney (having previously been for many years at Bowderdale in Wasdale) to the Gill, Kinniside, and one of the several Tyson families moved from Penny Hill, Eskdale to Farthwaite, Kinniside.

There were also opportunities for some families to spread out. Most notable in this respect were two of the leading Herdwick families, the Wilsons and the Richardsons. The Wilsons had had strong connections with Wasdale for about 300 years. John Wilson, the father of William Wilson, the HSBA secretary, had been born at Wood How, Wasdale, in the 1880s and later farmed Middle Row at Wasdale Head. His son, William Wilson subsequently moved to Watendlath where in 1920 he was described as having an 800 strong flock which had been "carefully handled" by him and his late uncle for 46 years. In 1924 he had moved to Stoneycroft, Newlands, though he also had Ashness in hand. In 1931 he moved to what he later named Herdwick View near Bassenthwaite Lake. His son, another William but more often known as Billy, remembers as a boy walking the surplus sheep from Ashness to their new fell, Binsey. The Wilsons were highly competitive as well as successful in the show ring and their tups were very influential in the breed for some years.

Herdwick Billy's brother, Richard M. Wilson, at the onset of the Association was at Middle Row, Wasdale Head, with his 600 strong "noted Ram Breeding Flock." Dick moved to Watendlath to farm with his brother for a while before going back to Wasdale for three years. In 1927 he moved to one of Herdwick country's biggest farms, Glencoyne, with its large amount of enclosed ground as well as open fell.[62] As already mentioned 500 of the Wasdale ewe flock were left at Middle Row where they were taken over by Joseph Naylor from Caldbeck. Although there are still Naylors at Wasdale Head, there are no longer Wilsons at Glencoyne. Indeed years before they retired from there they gave up breeding Herdwicks although they did for many years buy pens of Herdwick draft ewes. But 'Young' Dick Wilson of Glenridding kept up the Wilson tradition by breeding fine Herdwicks until the mid-1990s.

The Richardsons were another prominent family in the Herdwick world. The Richardson dynasty of Herdwick sheep breeders was started by John Richardson who farmed at Clappersgate before moving to Seathwaite, Borrowdale in 1885. The 1879 *Shepherd's Guide* has an entry under Ambleside, Rydal and Loughrigg for John Richardson indicating that his fell sheep "Go in Wythburn Head." John was twice married and fathered seventeen children. R. H. Lamb described John Richardson as, "one of the most famous breeders of his day" with his sheepwalks extending up to Great Gable and almost constituting "the hub of the breeding ground."[63]

He died in 1915 and the family relinquished the tenancy at Seathwaite, but several of his sons went on to farm on their own account. For instance, a son from his first marriage, Joseph, farmed at Gillerthwaite, in Ennerdale, before moving to the great Kinniside farm of Swinside. Joseph's son John farmed there until the 1960s and had a good deal of success in the show ring. Another grandson of the elder John Richardson, William, farmed at Fell Side, Bootle, until his death in the 1990s. Another son from John (the elder's) first marriage, William, farmed at Swinside in Thwaites parish before moving to Eskmeals. He carried on keeping Herdwick sheep, descendants of some

Gable Blueboy with Jerry Richardson and sons winning the Keswick May Fair, 1937.

ewes taken with him from Seathwaite, even when he later farmed in the lowlands at Silloth and at Armathwaite.

Jeremiah (Jerry – the eldest son from the second marriage) was born in 1882 at Clappersgate, just before the move to Seathwaite. On the death of his father and before the arrival to Seathwaite of the Edmondson family from Birkrigg, in Newlands, in 1916, Jerry and his wife Grace set up in their own right at Pike Side, Ulpha, in 1915, where sons Thomas and John were born. He took 60 ewes from Seathwaite with him to Pikeside along with the flock prefix 'Gable'. The flock prefix and the sheep went with him wherever he was and the sheep were retained for many years as a separate stock.[64] In 1920 he moved to Low Snab, Newlands, where Joseph (Jos) and Betty were born. According to Betty the move to the Snab was a "bad move" – the house was in a poor state of repair and the move coincided with the collapse of sheep prices when the post-war economic boom finally evaporated. Such was the available mobility however, that the family moved again to one of the Watendlath farms next to the Tysons.

They were to spend ten years there, before moving in 1934 to the much larger enterprise at Gatesgarth, which they hoped would provide opportunities for the children as they reached maturity. Jerry died in 1953 after which Thomas took over the running of the farm with assistance from his brother Jos and from George Birkett, his brother-in-law who had married Betty. George Birkett was from Gillerthwaite, the dalehead farm in Ennerdale. The Birketts stayed at Gatesgarth until 1958 when they got a farm of their own in Langdale, and later took on the tenancy at Tilberthwaite at Coniston. Brother to Betty, and to Thomas, and Jos, was Johnny Richardson. During the war he had been a prisoner of war of the Italians from whom he escaped, living clandestinely for some time with an Italian family. After the war he became huntsman for the Blencathra Foxhounds but also did a great deal of work amongst sheep during his summers. He died in 1986. Thomas's son Willie continues to farm Herdwick sheep at Gatesgarth.

Isaac, the youngest from John Richardson's second marriage, had three of the best Seathwaite ewes which were registered as Herdwick flock number 2, "the best blood in the Herdwick world." He farmed at Middle Row, Threlkeld, from 1924 to 1936, and then moved on successively via Thornholme, near Calder Bridge, to Kiln How in Borrowdale, on to a farm at Waberthwaite, before ending up farming in the Eden Valley at Kirkoswald. He retired to Threlkeld. Simpson was at Taw House, Eskdale, before moving in 1932 to High Lodore, in Borrowdale. Although he moved 50 miles by road from his previous farm, the sheep from the two farms were fairly close neighbours at the fell. The trajectory of these sons of John Richardson also, of course, illustrates the sheer mobility of fell farmers between the two world wars. After the war, and with the arrival of state support for hill farming, and a presumption towards expansion, things became more settled and often farmers were able to buy their farms.

Agreeing Breed Standards

Right from the outset of the Association there was debate about standards. At a meeting of the Executive Council at Keswick in August 1920, "Questions were asked regarding Flock No 109." This flock belonged to Joseph Bellas, Souterfell, in Mungrisdale. In the first *Flock Book* it was stated that Bellas put 200 ewes to the tup and that it was "a flock of pure Herdwicks, of old

---

Standards of Excellence and Value of Points

HEAD (Points 12)

Rams To be masculine in character, face of medium length, broad and full between the eyes muzzle strong, nostrils, wide open; strong jaw; ears of medium length, white and alert. Eye, prominent and bright. Face, jaws and top of head covered with strong, bristly hair and free from wool.

Horns smooth and round, whitish in colour, low set, wide apart and rising well out of the back of the head.

Ewes to correspond, except that the head should show a distinctly feminine character and be entirely free from any sign of horns.

NECK (Points 6)

Medium length, rising well out of the top of the shoulder which should be well laid back and deep.

BODY (Points 25)

Ribs well sprung, well-filled behind the shoulder. Back broad and flat, strong loins. Hind quarters set squarely on good legs of mutton well fleshed to the hock. Breast deep and prominent.

LEGS (Points 10)

Fore legs straight and clean with big knees and springy fetlocks. Hind legs strong, flat boned, straight and well covered with strong, bristly hair free from wool. All four legs set well outside the body, and big white hoofs preferred.

TAIL (Points 2)

Thick, strong and full of muscle.

UNDERSIDE (Points 5)

Sheep when turned should have a deep chest and prominent breast bone. Broad across the belly and well covered with wool of a clear and white or blue colour. The hind legs should be straight and lie flat out on the ground.

COAT or WOOL (Points 25)

The coat should be heavy and dense with a good undercoat of fine wool of even colour and quality over the whole of the body, with a stronger ruffle or mane round the neck and top of the shoulder.

COLOUR (Points 15)

The face and legs should be a clear "hoar-frosted" colour on all sheep.

TOTAL POINTS = 100

OBJECTIONS

Black-spotted face or legs.
Brown or Yellow colour in any part.

standing." After considerable discussion it was agreed that the owner be asked to gather the flock for inspection. An inspection took place, but it is not possible to discover what action, if any, was taken. What we do know, however, is that the Council in November 1920 discussed 'fixing type'. The record reveals that, "the question of fixing a type to breed to, was gone into: after considerable discussion and many suggestions it was finally decided to drop the matter for the present."

At the 1921 Annual General Meeting a few months later, however, there was "a long and interesting debate" on the same subject. This provoked "many different opinions" and led to the appointment of a twenty strong committee to go into the subject. The committee met a month later on 19 February 1921 in Keswick for the purpose of "discussing and settling the standard of excellence." Again there was "considerable discussion," but the outcome of the meeting was agreement, with breed standards being drawn up with great skill by S. D. Stanley Dodgson. It was not long before the new standards were being tried out. At a meeting of the May Fair Committee just before the Keswick Tup Fair May Meeting in 1922, Stanley Dodgson proposed and William Wilson seconded – and it was agreed – that "the Judges each to be given a list of Points and to keep as near to them as possible."[65]

A more important consideration as time went on, was that numbers of registered tups declined markedly during the 1930s. Initially there had been an annual average registration of about 450 tups per year, but this declined to 182 in the 1935-36 *Flock Book*. In seeking in 1936 to answer the question why this had happened, R. H. Lamb felt that it was because the registered ram did not live up to expectations. There was just a small fee for registration per ram (two shillings and six pence – 12.5p in today's money) and members themselves decided what they wanted to register. New member's flocks were subject to only a perfunctory viewing. Mr. Lamb commented that rams were registered freely, "and with such lack of discrimination that as time went on it became more and more painfully apparent that a ram with a handle to its name was worth no more than the commoner."

In 1929 Frank Fawcett (who shortly afterwards became President) advocated an organised scheme of inspection for individual merit before registration. Shortly afterwards R. H. Lamb became secretary and he worked with Fawcett to ensure the introduction of the scheme. The registration scheme was based on 'inspection for individual merit'. From 1930 all tups had to be passed as Herdwicks 'of reasonably good merit' by a panel of three inspectors who were also empowered to reject inferior animals. Fawcett firmly believed, however, that the only way to drive up quality further still was to set up an inspection scheme for ewes. Only registered ewes would be allowed to produce tups eligible for registration.

At the Annual General Meeting held at the Black Lion, Whitehaven, on 4 February 1931, it was decided amongst other things to revise the rules and that, "some form of ewe inspection be considered." This was left to the

The Ewe Inspectors: Wm. Rawling, J. N. Gregg and Jos. Plaskett.

Council to devise with the recommendations to go to an Extraordinary General Meeting. The latter took place on 2 March 1931 at Cockermouth and was attended by twenty people.[66] Jonathan Rawling moved, and Wm. Rigg seconded, that the council's scheme be adopted. This involved on-farm inspection of ewes with their lambs at foot. Jerry Richardson and Jos. Richardson proposed that it be rejected. An eleven to eight majority narrowly supported the original motion. Jonathan Rawling's suggestion that the ram inspectors should also serve as ewe inspectors was, however, supported unanimously.

The scheme clearly caused great consternation and after lambing time was largely over, on 11 May 1931, another Extraordinary General Meeting was held attracting 22 members on this occasion. Frank Fawcett from the chair explained that the meeting had been called on account of the, "currency of dissatisfaction and erroneous impressions" regarding the scheme. After a discussion and resolutions for and against a narrow majority was found for the original scheme, with an additional point being agreed, i.e. that ewes should be inspected in September along with the rams. The scheme got underway but a significant number of noted breeders boycotted the scheme. Fawcett wished to push further forward, however, and at his suggestion a system of ewe pedigrees was promoted whereby only a percentage of each flock's ewes was eligible for registration. Only the offspring of these ewes would be eligible for inspection as shearlings.

This more demanding ewe inspection system was agreed to (though never with a strong majority) and in 1934 three inspectors (J. N. Gregg, W. Rawling and Jos. Plaskett) toured all those farms that were prepared to participate. The inspectors had the power to register ten per cent of a member's flock. The inspectors registered 1,300 ewes and "only the progeny of them are eligible for inspection" – though the individual animal still had to have merit in its own right.[67] That is to say, it was not automatically the case that the male product of the mating of a registered tup with a registered ewe was a tup lamb acceptable a year later as a registered tup.

There is very little doubt that this scheme was highly controversial in the Association and that it experienced a fairly high rate of boycott. To be sure the onset of the ewe inspection scheme accompanied the Great Depression of the 1930s with R. H. Lamb recording in an editorial in the 1931-32 *Flock Book* that, "So dark indeed was the outlook in the autumn of 1932 – that it would not have been surprising if interest in the Association had waned." But even without the impact of the depression it was clear that the ewe inspection scheme was not working. Two years later in the 1933-34 *Flock Book*, Lamb suggested that, "the full adoption of the scheme might have brought almost full control of the breed."

There was clearly some recognition that the ewe inspection scheme, as good as it ideally was, did not suit the temperament of many of the Herdwick breeders. It involved too much regulation and 'red tape'. It also offended against principles that many of the breeders held dear, especially the importance of breeding sheep that were fit to do their job on the fells. To this day there remains a belief amongst many breeders that it is important for a good proportion of tups to be bred from the fell stocks rather than just from inevitably 'kin-bred' small elite flocks of tup-producing ewes. At a meeting of the Association's council on 13 April, 1936 Thomas Bainbridge of Borrowdale and J. F. Buntin of Great Langdale moved and seconded that, "the registration of gimmer sheep be optional, and that rams and gimmers from unregistered sheep be eligible to come before the inspectors for the year 1936." This was confirmed unanimously at an extraordinary general meeting at Keswick a month later. From then on effectively the ewe inspection system, though it did have its adherents, eventually died out quietly.

Losing Ground

The general economic depression of the early 1930s took its toll in a general way as well, and widened the Association's essentially internal concerns such as breed standards to include concerns about the viability of fell farming generally. The well-informed local writer, E. M. Ward, based at Grasmere, recalled the 1930s on the Herdwick sheep farms as being a time, "when sheep prices collapsed, Herdwick lambs were sold for little more than cock chickens, and farmers were enabled to carry on only by turning their homes into boarding-houses for summer visitors."[68] Similarly, Mrs. Heelis wrote to

Eleanor Rawnsley on 24 October 1934 bemoaning the fact that her tenants at Yew Tree, Coniston, had not been able, due to illness, to open their tearoom that summer. She noted that, "the farm can scarcely pay without the teas and visitors."[69] There was very little money about and people inevitably dispensed with the optional extras.

The arrival of war in 1939 improved matters, though it also created its own problems such as the call-up for military service of young fell shepherds and the loss of hogg wintering ground as a large amount of grazing land in both the fells and the lowlands was ploughed out for cropping. There was a high level of state control and forceful direction, not always popular but usually effective, of local farming through the War Agricultural Executive Committees. Prices were taken under Government control and improved markedly. But grain, potatoes and milk were the priorities and sheep numbers fell by 30% nationally.[70] There were, however, incentives to breed sheep in the hills – with a hill subsidy coming in which the HSBA wished to see extended to cover gimmer hoggs.[71]

After the war problems with shortage of hogg wintering persisted, notably on Walney Island where the Butterilket hoggs had habitually gone. But generally speaking things were on the up with the post-war Labour Government introducing the Hill Farming Act of 1946 bringing with it not only the beginnings of peacetime agricultural subsidy but also the beginnings of peacetime regulation. There was significant opportunity and need for the Association to lobby for the interests of the fell farmer when things might be heading in the wrong direction. The Association retained its concerns (at the 1949 AGM) about the potential establishment of a national park in the Lake District because of potential problems of damage to walls and fences, gates being left open and stock straying.

Nothing could be done, however, about the weather. In May 1946 the Association's Annual General Meeting was marked by complaints from fell flockmasters about the heavy losses of sheep there had been over the winter and eighteen months later in December 1947 the losses of that year were reported as having been even higher. In the winter of 1946-1947, the snow fell in early February 1947 and stayed, frozen on the fells, until the spring – bringing about huge losses of fell sheep and a very small crop of lambs. A contemporary estimate of losses was 40% of breeding ewes.[72]

But the biggest challenge to Herdwick sheep in the 1940s was probably not the weather but other breeds of sheep. The Rough Fell was taking over in some parts of south Westmorland, but the biggest loss of ground was to Swaledale sheep in the Caldbeck area, back of Skiddaw. Also Keskadale in Newlands had become a Swaledale farm as had several in the Hartsop area.[73] Ward reported that, "Sheep of a Swaledale-Herdwick cross now wander over the top of Helvellyn." There was also a great deal of cross-breeding in both the Buttermere and Coniston districts in an attempt to get – as well as the inevitable hybrid vigour – a bigger sheep, with, to some extent, better value wool. There were even considerable numbers of Swaledale tups at the Fell Dales Association show at Eskdale for a number of years.[74] The Keswick May Fair Committee had, as early as 1935, discussed a suggestion that all Swaledale rams should be penned on a different side of the field to the Herdwicks – a sign almost certainly of the opposition that some breeders had to the growing number of Swaledales at the major spring event in the Herdwick world.[75]

Farmers who favoured Swaledales were busy creating their own institutions such as the Mungrisdale Swaledale Ram Show which had its 77th show in 2007. The Association's secretary, R. H. Lamb, before his premature death at the end of 1943, was fully aware of the dangers to the Herdwick breed from the Swaledale and had advocated trying to breed bigger Herdwicks for commercial considerations. He had noted that since the early 1920s there had been as many horned sheep as Herdwicks at the annual Wyllie Gill shepherds' meeting at the Back of Skiddaw.[76] Even eminent Herdwick breeders were tempted to increase their returns by using Swaledale tups to an extent as part of their cross-breeding programme. For example, when in the early 1930s William Rawling (who ended up farming at Godferhead, Loweswater, and who was one of the Association's ewe inspectors), moved from the Hollins at Ennerdale to Little Braithwaite, he had a sale of 255 sheep included amongst them being 95 Swaledale and Swaledale-Herdwick cross ewes and shearlings alongside 70 Herdwick ewes. There were also 24 grey-faced gimmers and 22 Swaledale-Herdwick hoggs alongside the Wensleydale, Leicester and Border Leicester tups which no doubt sired the grey-faces. At the same sale his brother Jonathan, who was moving onto the Hollins, put onto the market surplus sheep from his tenancy at Black How, Cleator. Amongst the 292 sheep available ("direct off the fell," Dent) were 160 Herdwick young stock sheep in lamb to the Wensleydale and 40 Herdwick shearlings which had been put to the Swaledale tup alongside 40 Swaledale and Herdwick wether hoggs.[77]

An academic volume on sheep published in 1945, however, still largely classified the Swaledale as a northern Pennines breed numerically far less important than the Scotch Blackface "though it is not losing ground." The same source indicated that the Herdwick – "the breed of the Lake District" – was still, "of considerable numerical importance."[78] The truth of the matter was that the Herdwick was losing ground to the Swaledale.

2: Fells and Heafs

"Twelfth and thirteenth century sheep grazed on the mountains as gladly as their present day successors, and in those parts at least the aspect of the land has changed little. The sheep walks of the Cumberland hills are as they were when William I became king and as they had been from the remote past." Doris Mary Stenton,
English Society in the Early Middle Ages (1964)

Heaf-Going Sheep, Fell-Going Sheep

Although the word 'Herdwyck' – meaning sheep pasture – is recorded in documents going back to the sixteenth century, this does not mean that the breed of sheep that acquired this name has remained unaltered over the centuries. What has remained fundamentally unchanged is the grazing of the open fells of what came to be called the Lake District by flocks of sheep. Records of this activity go back well before the sixteenth century. The area concerned consisted largely of open fell land, described as manorial waste and 'forest.'[1]

H. S. Cowper in his *Hawkshead: its history and archaeology* published in 1899 stated that, "Herdwick is really the name of a sheep farm not of the sheep... the name being applied to those farms on which the flocks were let with the land. This custom was peculiar to the Lake District of Cumberland, Westmorland and North Lonsdale, and the sheep which were thus let being of a peculiar and distinct breed, gradually and naturally assumed the name of the farms upon which they were bred, and became universally known as 'Herdwicks'."[2]

Garnett, after going through various descriptions of the breed from the nineteenth century, states that irrespective of origins or genetic purity, "there was a white or 'rimy' faced, hornless, short-wooled breed of sheep on the mountains of Cumberland and Westmorland from a very early date, which took their name from a mode of agriculture peculiar to the district – the farms on which they pastured being called 'herdwicks', i.e. sheep and lands were let together."[3]

We can only assume that the sheep that Furness Abbey kept in the thirteenth century on its vaccary or grange, in upper Eskdale at Brotherilkeld (Butterilket in dialect and from the Old Norse for the booth – summer dwelling – of Ulfkell[4]) were bred largely for their wool and were probably a far cry from the present breed – but they were being produced through the exploitation of the fell grazings. At this time keeping cattle was probably more important than keeping sheep, but in the records of the Commissioners of Henry VIII for 1537, included in the revenues for Furness Abbey, there is an item for "herdwycks and shepecots" amounting to £39-13s-4d. In the same document Sir J. Lamplugh pronounced that, "Erleghecote haythe always beyn a hyrdewyke or pasture ground for the schepe of the abbottes of Furnes."

West in his *Antiquities of Furness* in 1774 cites a decree of 1564 which talks about parcels of ground known as, "the Herdwick called Waterside Parke... the Herdwick called Lawson Park... [and] the Herdwick called Plumers."[5] A hundred and twenty years later Thomas Denton, in his *A Perambulation of Cumberland* of 1687-1688, recorded of Sir William Pennington of Muncaster that, "He hath an Herd-wick of sheep upon Mulcaster fell."[6] Similarly in his will drawn up in the 1670s, Robert Rawlinson of Grizedale Hall gives to his cousin, "All that Herdwick called Lawson Parke lying and being in the ffourness ffells in the County of Lancaster"[7] – we must assume that the sheep went with it too.

The foremost dialect dictionary from the mid nineteenth century confirms these usages. A 'Herdwick' was, "the mountain sheep of Cumberland." A 'fell sheep' was a synonym for a Herdwick. A 'heaf' was, "the part of the mountain or moor on which any flock is accustomed to depasture..." 'Heaf gangan' was the term used mainly in the central and western parts of Cumberland for the word 'hefted' (which was used more in the northern parts of the county) with the dictionary entry reading as follows, "Hefted sheep are mountain sheep let along with a farm and depastured on a particular part of the common or fell termed their HEAF."[8]

Landlords' Flocks

"It is a circumstance, perhaps, not generally known that a farmer of fell lands, rents likewise the sheep depastured thereon. Were that not the case, the sheep would 'mingle on the mountains,' and be a perpetual source of anxiety and trouble. Every fresh renter has the sheep numbered, and valued, and he conditions with his landlord, to leave upon the estate, when he quits it, the like number and value." Wm. Green, *The Tourist's New Guide to the Lakes,* 1819.

Fell land without sheep heafed to it was of little value and would prevent farms from changing hands advantageously. A deed concerning the transfer of parcels of land in the 1750s within a family called Jackson at Hazelhead, Ulpha, described not just the land but included its wider productive capacity, "all houses byars stables... moors mosses mountains sheepheaves common of pasture and turbary." Fifty years later the family was just as concerned to capture this wider resource. In 1801 William Jackson's property included two tenements at Hazelhead which included, "all buildings... fronts... folds dunghill steads gardens orchards... feedings sheapheath brackendales and woods." The sheep were not forgotten in all this. To a nephew, William, he left the farm at Holehouse, in Ulpha, "with the stock of sheep." To another nephew, Miles, he left, "the other half of my stocks of sheep." William left yet another nephew, John, "the freehold messuages which I purchased of my late brother Jonathan Jackson at Hazelhead, with half of my stock of sheep now let with my estate there."[9]

Landlords were naturally concerned that the fell farms they rented out had some rent-returning economic viability as a couple of examples from the late seventeenth century indicate. When in 1696 the then owner of the How (now called How Hall) in Ennerdale, Sir John Lowther, was considering selling it, William Gilpin, his agent, wrote to him of a proposal by a third party to buy the How in Ennerdale. Gilpin urged that, "A decision is urgently needed because if the flock it bears are 'dissipated' the buyer will withdraw."[10] In the mid-nineteenth century, when the Dickinson family owned How Hall, there was a flock of 440 sheep let to the tenants.[11]

Driven by similar reasoning, when Sir Daniel Fleming acquired Kentmere Hall, a process he had started in April 1698, he followed it up by ensuring that he acquired 67 ewes and lambs, one wether and twenty geld ewes in mid-June. This heafed stock would form the basis of the ability of the farm to exploit its share of the fell grazings. It would also, by putting sheep to the fell through the fell gate, enable the farmer to increase significantly his management of the in-bye, for instance, by being able to bar off fields for hay and arable crops.[12]

Rational management behaviour of this sort led to the development of the landlord's stock system where the tenant on entering the farm takes over the landlord's flock of sheep as well as any surplus sheep the outgoing tenant is prepared to sell him. This system recognises that these fell farms, which typically have less than 100 acres of better land, depend on the fact that they have a flock of sheep which knows where its fell is and which grazes it. If the fell flock is sold off from the farm, the farm's economic value is considerably reduced and it can take years of difficult and painstaking work before a fell-going flock can be properly re-established. The necessity for such a relationship between land and stock was recognised hundreds of years ago and the name seems to have migrated to the sheep. It had in fact for many years been common for details of the heafed flock – in short the Herdwicks and their management – to be stipulated in letting agreements. Garnett quotes an agreement dated 1741 concerning another Kentmere farm. Not only had the house, gates, hedges and fences all to be kept in good repair, and the manurial value of the farm be maintained, it was also necessary:

"to give sufficient security for the redelivering of 80 heaffe bred and heaf going sheep, upon his ancient Sheep heaffe at Over End... at the expiration of his Farm, of the like number, sorts and kinds and in such plight and condition in every respect to the judgement of four persons to be

Indifferently Chosen between them, as they shalle be delivered by the Landlord to the Tenant." [13]

The effective use of the fell was one of the factors which brought viability. Sand Ground at Hawkshead was, for instance, rented out by the local Poor Law authorities as a source of income in the 1780s. It has been well described as comprising house, buildings, enclosed land and common rights and, "its flock of Herdwick sheep that all the year round lived on a portion of the open heights they had won for themselves by driving intruders away from it. It was known as their 'heaf', and because they were strongly attached to it, and would give endless trouble if taken elsewhere, they were treated as part of the customary holding, and were let or sold with it."[14]

The landlord's stock was a precious asset and was valued accordingly often using a bond system, with money being lost by the tenant if at the end of the tenancy the farm had not maintained adequate quantity or quality in the flock. The bond for the landlord's flock at Glencoyne in 1824, for instance, was for the remarkable sum of £2,000. The landlord's flock at that date consisted of 214 sheep made up of 57 twinters (shearlings), 43 gimmer hoggs (ewe lambs), 43 wether hoggs and 71 older wethers. Not only did the landlord's flock system exist to secure the new tenant the ability to earn a living from a fell farm, it also sought to protect the heafing system itself.

Livering day at Middle Row, Threlkeld, 1980s. Viewers are (left to right): Johnny Richardson, Joe Cowman and Stan Edmondson.

Disruption to the heafing system could be caused by a tenant introducing non-heafed sheep onto the fell or selling off some of the heafed sheep to others, thus enabling them to gain a foothold. The documents for the Glencoyne sheep stock in 1824, for instance, stipulated that the tenant farmed, "without buying any to heath, depasture or go amongst them or selling any that may in anywise return to haunt, disturb, incumber, inter common, graze or go amongst them upon the said heaths, fells, moors or pastures and commons..."[15]

Under this system the incoming tenant would, however, usually have (and has today) the ability to buy any surplus stock on top of the landlord's flock if such were available, and could be afforded. This was one clear way to prevent surplus heafed sheep falling into the hands of others trying to get or extend a foothold on the fell. If the heafed sheep did not stay with the farm, they might be bought by a neighbour who would continue to graze them at the fell where, of course, they would stay on and effectively 'steal' their old heaf. There was also the very real problem of sheep owned by several separate people and being kept on the fell having the same earmarks. To this day, surplus sheep when being sold off a fell farm are often presented for sale in the auction ring with a firm statement from the vendor that the sheep are, "not to go back onto [such and such] common."

Agreements also specified other criteria of good husbandry of the fell flock. The tenancy agreement used for Pikeside, Ulpha, in 1926 seems to be (if the language is anything to go by) one of some antiquity. But the terms of the letting agreement had stood the test of time and still met contemporary needs. The agreement articulates the concepts of sustainable flock management that are so crucial for maintaining the assets of a fell farm:

"The tenant will draw the sheep according to the usage of the country but not so as to lessen the number of any sort – he shall take the fleeces in shepherdlike manner and keep each sheep marked with the particular ear and tar marks heretofore used. He shall use his utmost endeavours to maintain in a shepherdlike manner each sort undiminished in number and sound. In the last year of the tenancy he shall deliver up the stocks of sheep sound and equal in every way to those aforesaid according to the judgement of two indifferent persons, one to be chosen by the Landlord and the other by the Tenant. Stock of sheep – ewes 86, shearlings 15, gimmer hoggs 26."[16]

The incoming tenant's intention would be (and still is) to breed up a surplus (usually called an 'overplus' locally) that would belong to him. Typically a landlord's stock is only a proportion of the stock that the farm and its related sheep heaf on the fell could carry, so that conscientious shepherding can generate a surplus thus building capital in order to go onto a bigger farm or to be able to retire. The size of the landlord's flock is typically about

The Harrison family of Butterilket.

a quarter or a third of the farm's carrying capacity – though there are some farms with very large landlord's flocks such as Yew Tree at Rosthwaite, where it is 814 sheep, and Middlefell, Great Langdale, where it is 668[17] – circumstances which, not surprisingly, leave little scope for capital growth.

The National Trust is clearly conscious of the fact also that sometimes the landlord's flock is too small for an incoming tenant to build up from sufficiently quickly. At Stonethwaite, Borrowdale, for instance, on a recent change of tenancy the landlord's flock was increased from 66 sheep to 211 sheep, made up of 141 ewes, 35 shearlings and 35 gimmer hoggs.[18] When Seatoller was given to the National Trust in 1959 by the Treasury in lieu of death duties, it ensured that the farm remained as a working one by purchasing the landlord's stock which went with the farm using money from the Goodwin Fund.[19]

The key thing is that the landlord's stock is there from which the flock can be grown by the sustainable exploitation of the heaf. The turning over of the landlord's stock from an outgoing to an incoming tenant still takes the same form today as it has taken for hundreds of years – at the 'livering' or 'viewing' day with viewers acting on behalf of the different parties and trying to agree about the quality of the stock. The National Trust currently has a panel of three farmers, each of them Herdwick breeders, but who are not Trust tenants, to act on its behalf.[20] A key document is the 'Viewing Paper' which records the state of the sheep that are being taken over, with sheep being described by their 'number and kind' and their 'state and condition,' as for instance, in the document of 22 April 1986 when there was a change of tenancy at Tilberthwaite; 300 Ewes "fair average," 75 "gimmer twinters "fair average," and 100 gimmer hoggs "fair average."[21] Needless to say, these 'sheep viewings' or 'livering days' (when the sheep are 'delivered' from the outgoing to the incoming tenant) can sometimes be a relative formality as in the case above – and on others can present some difficulty if the sheep are not up to standard.

It is also important to note that the landlord's flock system gives the tenant a substantial leg-up as well as the ability to exploit the fell. A decent size landlord's flock means that the ingoing tenants have less money to find when they enter a farm. This has been appreciated throughout the ages, especially since the outgoing tenant is also trying to sell to the incoming tenant the surplus stock sheep at as high a price as possible due to the fact that the sheep are heafed and command a heafing charge also. As Wilson Fox put it in his report on Cumberland to the Royal Commission on Agriculture in 1895, the system was, "a great advantage to men with but little capital who want to take a hill farm." He quoted Joseph Harrison of Butterilket, where 1,200 sheep came with the farm and, "who began life as a shepherd," as saying, "If it had not been for the system of hiring sheep I could never have taken this farm."[22]

The key thing, as R. H. Lamb put it in the 1930s, is the link between sheep and the fells. He argued convincingly that:

"the name 'Herdwick' is not properly that of a variety of sheep, but it is an old word for a sheep farm where sheep and lands were let together. This practice, which is of a very early date, and is not known in any other part of this country, involves proceedings which are of equally ancient origin and are regularly carried out to-day just as they were two hundred years ago."[23]

Wool and Wethers

"Herdwick sheep are bred by the fell farmers for their wool – and for mutton also, for there is no better mutton in England." H. H. Symonds, *Afforestation in the Lake District,* (London 1936)

Much of the importance of the heafing system is due to the essentially open nature of the fells of Herdwick country and the inclination of Herdwick sheep to 'go to the fell' and to forage; i.e. not to sit at the fell bottom waiting to be fed. The sheep have a propensity to do this – and the reputation they have (sometimes unfairly) for being prone to 'ratch' goes side by side with their ability to occupy the fell. A key role in this job of occupying the fell was often done by the keeping of a proportion of wether sheep in the flock. Mutton from four or five year old wethers was a major enterprise on the fell farms within living memory. Wethers grazed at the outer edges of the heaf and helped to keep the fell free from encroachment by other sheep and acted as a strong physical presence, discouraging incursion by sheep from other flocks.[24] Having a proportion of wethers in the landlord's flock might well have been crucial to keeping the heaf when a farm with only a small landlord's flock changed hands and there was the potential for encroachment by other flocks. A Great Langdale farmer, Joe Gregg, used to say, "A wall of teeth round the heaf is better than a wall of stone."[25]

Almost certainly the most important reason for keeping wethers was that they yielded an annually sustained and sustainable crop of wool at a time when wool was a valuable commodity – though its value fluctuated markedly. The Victorian writer, Elizabeth Gaskell, attending a clipping day (she felt that it was "a sort of rural Olympics") near Keswick in 1853, noted that the price at the time was twelve shillings (55p) a stone, whereas a few years earlier it had been worth up to £1 per stone.[26] Very great store was put on the harvesting of the wool crop with clipping days being days of great occasion and mutually collaborative effort – with the farms in an area having their regular clipping days or if the weather was not co-operative there was an order in which the flocks were clipped.

Thomas Bowes of Fenwick, Thwaites, (like many another fell farmer) clipped his sheep, year after year, at more or less the same time. The participation in the Fenwick clipping of William Fisher, who farmed at Barrow-in-Furness and who was Mrs Bowes's brother, was recorded in the latter's diary over many years: 8 July in 1820, 13 July in 1824, 9 July in 1836, 8 July in 1837 ("Thos Bows sheep shearing Henry and myself went and found all well"), 7 July in 1838, 9 July in 1839. The last mention of the Fenwick

Gillerthwaite clipping c. 1900: the men, note the fiddler and the dogs. Tommy Dobson, the famous huntsman is second from the right on the front row.

clipping was on 14 July 1855: "Son Richard with his Drs. Margaret & Eliza and son Wm. went to Fenwick to Thos Bowes Clipping returned on the 15th Do."[27]

Clipping days took on an important social as well as economic function and there was often substantial entertainment. Local newspapers often reported on the day's events.[28] The clipping at Gillerthwaite in Ennerdale was, for example, reported in the *Whitehaven News* for many years. The report of it in the issue of 20 July 1899 indicated that it was a three day event and involved 24 clippers (not just from the environs of Ennerdale but also from Wasdale and Lorton) and eight other men who acted as catchers, fleecers and smitters. There was a contingent of sixteen women, eleven of them unmarried, to provide food and other refreshments. There was singing and dancing in the evenings. On the final day, the proceedings concluded with a number of the men engaging in a wrestling competition. The wool was stored in the buildings – and kept until it was sold to wool merchants.

From about 100 years ago an additional reason for prompt and systematic summer gathering – and immediately thereafter clipping – emerged. About this time blowfly strike seems to have increased markedly.[29] In the years before effective chemical protection through dipping against blowfly strike it was very common for sheep with the wool still on them in mid and late summer to be at risk of being eaten alive by maggots.[30] This required shepherds to spend many hours at the fell, especially on humid days, with a pair of shears and a pot of tar, looking for struck sheep. The writer W. T. Palmer (who had been a Cumberland shepherd when young) wrote in 1905 that, "Instead of the shepherd piping and watching the sheep with lambs by their sides streaming over green swelling hills, in the English mountain-land it is

Gillerthwaite clipping c. 1900: the ladies.

the season of the detested maggot... when the shepherd should theoretically be at ease, he is really, ointment pot in hand, climbing about the roughest part of his holding."[31]

Clipping sheep on an as-and-when, emergency basis was vital if the 'mawk' or 'wicks' were about. Canon Rawnsley wrote in 1906 about West Head's clipping day (the second Monday in July), when friends and neighbours turned up to clip (having the favour repaid on their own farm on a reciprocal basis) as being able to be kept because, "as the 'fly' has not been troublesome this year, the day holds." He added that, "There have been years when, because of the 'mawk', they have had to forestall the date, and clip the flocks by instalments."[32] Normally, however, clipping day was a great occasion of communal work with each farm in an area having its own day. The 12 July was the day for Chapel Hill in Mardale. A contemporary observer wrote, "The sheep were got off the fell early in the morning in readiness for the shearers who came from everywhere around; and a pleasant picture they made, each with a sheep stool on which was bound a bonnie Herdwick."[33]

The Herdwick Sheep Breeders' Association's minute books are full of references to campaigns to tackle the blowfly in the years when, although salving (smearing a tar-based mixture between openings of the wool onto the skin) had been replaced by dipping, many dips were ineffectual. Other things were tried. For instance, in 1919 a stock of fly traps was acquired for use by the members. Virtually twenty years later, in 1938, the same problem was being discussed. A veterinary expert was called in to give lectures at Cockermouth and Broughton-in-Furness – and there was some discussion of developing a scheme involving shepherds being paid to place and look after the traps. The real breakthrough did not come until 1946 when the chemical DDT was used to bait fly traps in an experiment at Buttermere and Caldbeck. As local writer Graham Sutton put it, the traps were 100% successful "not one sheep wick'd." DDT was at this point not legal but DDT-based dips eventually became available and were widely used for several decades – with disastrous consequences for wildlife.[34]

Wool prices were very variable. In the First World War Herdwick wool had made up to one shilling and six pence (7.5p) per pound. But the inter-war years saw poor prices and at times wool was withheld from the market as a protest. There is a well-known story that one of the biggest flockmasters, Isaac Thompson of West Head, Thirlmere, in the early 1930s kept his wool clip for three years because it was only being valued at three pence a pound. Things did not improve and in the end he sold it all for two pence a pound. By 1936 things had improved and wool was making about seven pence per pound.[35] But even in these improving years and for several decades beyond that nothing was wasted or disregarded: 'daggings' (soiled wool clipped from sheep tails) would be collected and wool would even be plucked from dead sheep.[36]

Sometimes the proportion of wethers in the landlord's flock was very high: at Glencoyne, for instance, in 1824 there were 114 wethers as against 100 female sheep.[37] But normally there was a much greater proportion of female sheep. For instance, when the Leconfield Cumberland estate was sold in 1957, the landlord's stock which went on High Fell at Wasdale, consisted of 209 ewes, 67 gimmer shearlings and 37 wether shearlings, 84 gimmer hoggs, 62 wether hoggs plus four tups; the landlord's flock at Seatoller consisted of 273 ewes, 31 gimmer shearlings, 62 gimmer hoggs plus 47 wether hoggs; and Wasdale Head Hall had a landlord's flock on Wasdale screes which consisted of 133 ewes, 37 two and three shear wethers, 26 gimmer shearlings, 24 wether shearlings, 36 gimmer hoggs and 30 wether hoggs plus three tups.[38]

In more recent years wether flocks have almost entirely disappeared – partly as wool became an increasingly minor source of income and it began to make more sense to replace wethers with ewes. Accompanying this development came the selling of wether lambs as stores for lowland farmers to fatten as hoggets. W. T. Lawrence of the Cumberland and Westmorland agricultural school at Newton Rigg, as early as 1910, had carried out a trial on fattening Herdwick wether lambs and selling them at a good weight (average live weight of 83lbs) at the end of April for a good profit.[39]

Farming the Fells
"The sheep farmers and shepherds of the central fells are united by their very tasks and toils carrying them up on to the 'tops' and, often enough, down into the valleys on the other side. If their sheep meet on a ridge as, say, those of Great Langdale and Eskdale meet on Crinkle Crags and Bowfell, they share common interests and anxieties and are true neighbours, separated though their farms may be by a couple of thousand feet of mountain." B. L. Thompson, *The Lake District and the National Trust*, (Kendal, 1946)

John Sawrey, who farmed at Grassguards, Ulpha, from the 1890s, like many another fell farmer kept a diary in which he noted the dates at which tups were loosed, ewes tupped, cows bulled, pigs killed, foxes hunted, hoggs sent away to winter, hoggs returned from winter, fields ploughed, peats cut and stacked, lambs marked, hay made, turnips thinned and much else. He also regularly recorded the dates on which he washed sheep and on which he finished clipping: for instance, in 1894 washing finished on 16 June and clipping finished on 8 July.[40]

Sawrey also kept the weights of his wool sheets (note these old fashioned sheets were bigger than the present day ones which only contain about 50 fleeces – these would contain more like 200 fleeces) each usually weighed about 30 stone and his clip was about 150 stone or 2,100lbs, assuming a fleece weight on average of about three pounds this means that Sawrey clipped about 700 sheep. Although each year has its individual incidents and occurrences, essentially the same pattern repeats itself annually – with those two great changing yet unchanging variants, the seasons and the weather, being the great determinants of what goes on. Farming, and not least fell farming, is a fundamentally repetitive business. It has great 'time depth' – though it also exhibits change over time.

Fell farming is an ancient farming system. As Angus Winchester has shown, its fundamental pattern in the Lake District was established from the Middle Ages. By 1300 most Lakeland valleys had populous communities of tenant farmers and what had been summer grazing areas were being turned into permanent settlements. The lords of the manor controlled the dalehead areas, and ran substantial cattle farms ('vaccaries'). A vaccary was first recorded at Gatesgarth in Buttermere in 1267. At Stonethwaite, Borrowdale, there was one controlled by Fountains Abbey by 1302, and one of Furness Abbey at Butterilket in Eskdale, by 1292. Vaccaries were also recorded at Ennerdale and at Wasdale Head in 1322.[41]

For the next 250 years a key change was the dividing of these large manorial areas into smaller units which formed farms in their own right and which were tenanted. Within this process there was much informal, opportunistic, enclosure of common land to expand the grazing area under the farmer's control on their farm's in-bye land. Winchester mentions that this involved wholesale enclosure of lowland moors and, "piecemeal intaking round the margins of existing settlement."[42] Wordsworth characterised this as, "a numerous body of Dalesmen creeping into possession of their homesteads, their little crofts, their mountain-enclosures."[43]

Both the dalehead farms and the smaller farms would typically exploit the common or waste land away from the inbye land with complex systems of regulation being established as to sheep heafs and access points to the fell. There were, for instance, seven sheep heafs on the wastes at Wasdale Head by 1664 and Wythburn fell is recorded as having ten 'steads' by 1606. These sheep heafs were probably shared by groups of farms.[44]

Flock sizes, Winchester tells us, varied widely. He has analysed wills from Above Derwent between 1566 and 1590 and has found that all the farms had cattle (clearly indicating the importance of milk as a source of food in the domestic economy), and 90 per cent had sheep. Sixty per cent of the farms had fewer than 100 sheep; 25 per cent had over 150 sheep. He also points out that there were, in the Lake District generally, some very large flocks indeed – with over 4,000 sheep being summered at the top of Kentmere in the early seventeenth century and flocks of over 100 sheep on farms of only twenty acres. He cites the contemporary observer, Thomas Denton, in 1687 writing of great mountain flocks at Mungrisdale, Lamplugh, Lorton, Uldale and Borrowdale.[45]

Accounts of the Lake District fell parishes in the late eighteenth and early nineteenth centuries also bear out the substantial scale of sheep-keeping on the fells. In Hutchinson's *History of the County of Cumberland* published in 1797, a report from a correspondent describes Brotherilkeld as, "a sheep farm of prodigious extent" and it was noted that there were about 14,000 sheep kept in Eskdale, Nether Wasdale and Wasdale Head. Of the sheep it was said that, "they are continued of the ancient breed, and small." (As R. H. Lamb pointed out there were other references for other parishes to the native breed and the home breeds. He searched without success for the name Herdwick in Hutchinson and other reliable sources, but he is quite confident in saying that the sheep referred to were the Herdwicks of the day, the fell-going sheep.)[46]

William Green in his *The Tourist's New Guide* to the Lakes of 1819 gave a comprehensive list of the large flocks of the Lake District, from which a few of the examples which follow are taken. He began by commenting that he did not think that there was anywhere else in the district where as many sheep were kept in one small area, as there were on the adjoining farms of Toes (Taw House), Brotherill Keld (Butterilket) and Black Hall. William Tyson of Black Hall had over 2,000 sheep, Joseph Rogers, the tenant at Butterilket (1,400 acres), had a flock of 3,000 sheep and Thomas Towers, the owner of Taw House, had something under 2,000. There were flocks of over 1,000 and nearly 4,000 sheep at Lorton, Forest Hall, Thornthwaite Hall, Shap and Shap Abbey – though interestingly these sheep flocks were owned rather than rented.

Green also gave a lengthy list of farms where the sheep were rented with the land, listing amongst them, to give but a few examples, Patterdale Hall with its flock of 1,700, Glen Coin [Glencoyne] in Patterdale with its of 900, Troutbeck Park with 1,500, Gatesgarth in Buttermere with 1,300. John Wilson had 1,000 at Rosthwaite and Thomas Wilson had "upwards of 1000" at Watendlath. There were 800 at Hartsop Hall and a similar number at Wilson Stable's Wasdale Head farm and also at George Martin's Tilberthwaite. Green wrote, "These are the greatest shepherds amongst the Northern mountains."[47] These farms are still continuing their business as large fell farms, many of them running Herdwick sheep flocks.

The keeping of such substantial flocks, alongside many dozens of smaller ones, would not have been possible without the organised and systematic exploitation of the fells. Access to hill grazings – largely common land – was a complex matter which was regulated by the manorial court system, which in part included elements of self-regulation. But there were clear rules and conventions as to the numbers and types of sheep and of other livestock which could be depastured on the fells at particular times. Needless to say disputes were frequent and legion, as the records of the manorial courts indicate. The greatest sin in the book (other than sheep stealing) was 'hounding' or 'dogging' of other people's sheep.

As Winchester puts it, "the frequency of presentments against individuals disturbing, chasing or hounding their neighbours' stock suggest that the herding of livestock on the open pastures was a perennial source of potential conflict between neighbours, notwithstanding the regulatory framework."[48] He also devotes a whole chapter to the subject of 'Good Neighbourhood on the Common Grazings'. An important issue here was not just that each flock had its own heaf, but also that documents record the detail of the 'rakes' and 'drifts' – i.e. the routes by which the sheep could be driven to and gathered from their heafs. The communal organisation of the exploitation of the fells was set against a regulatory framework established by the manorial courts governing, "the day-to-day management of the hill pastures." It is, of course, the case that whilst 'good neighbourhood' was the ideal, there were plenty of examples of 'bad neighbourhood' – sometimes leading to violence and injury and, on rare occasions, death.[49]

F. W. Garnett in 1912 in his *Westmorland Agriculture* quotes part of an unidentified dialect poem which illustrates the potential of common land grazing systems to create friction:

"We've fratched and scaulded lang and sair, about our reights on't fell,
The number of our sheep, and whaur the heaf was they sud dwell…
And oft we fratched and fret about, and throppled uddar sair,
Upon the whol' the fell hes meade mischief for iver mair."

The fell may have caused mischief, but it also supported sheep and thereby people. Communal resources are notoriously difficult to manage as the records of manorial courts bear witness over many years, although it must be said that manorial records are more likely to indicate when things went wrong rather than when they went right. People have, however, got by, and often more than a semblance of collective management of the grazing resource has prevailed. This usually depended over the years on a 'give and take' attitude being adopted.

A. W. Rumney wrote an eloquent and convincing account around 1912 of fell farming in the Newlands valley, which covers these issues in an authentic manner. Rumney wrote that his book did not, "aspire to the status of a novel" and he believed it to be, "the only attempt to describe a dales farmer's life and attitude to the world from the inside." When his hero took over his farm, a neighbour showed him where his 'fell' was even though there was no fence to prevent his sheep, "travelling anywhere on the eighteen miles' long stretch of fells that surrounded the sides of the dale, and only in one place to keep them from the common of the adjoining vales."

In a chapter called 'The Politics of the Fells' he outlines the story of "an unceasing warfare" between the farmers of Nardale (Newlands) and those of Scardale (Borrowdale), "each endeavouring to get the little bit of overlap, and increase their flocks, a warfare for the most part carried on in one another's absence, but once in a decade or so resulting in collision. Hasty words, an angry blow, or even a regular set-to on the debatable ground." Rumney's hero is gradually introduced to the culture and conventions of being a fell farmer. He learns about landlords' flocks, sheep identification marks, the role of four-year-old wethers in keeping the heaf, the occurrences of sheep stealing, the code of conduct of how encroaching sheep might be 'hounded off.' The 'bone of contention', however, was the watershed, with sheep constantly in some cases being shepherded onto the heafs of neighbouring farms.[50] But generally, life and work proceeded – lambs were heafed, sheep were sheared, wethers were sold and some sort of livelihood was generally made.

Shepherd's Guides and Shepherds' Meets

"Should anyone be found keeping a sheep, the property of another person who has entered in this book, for a longer time than is necessary for making application to some one having a book, (should he not have one himself,) it will be considered that he intended to defraud the owner of it, and will be prosecuted according to law." William Hodgson, *The Shepherd's Guide or a delineation of the wool and ear marks of the different stocks of sheep in Lancashire, Cumberland, and Westmorland,* (Ulverston, 1849)

Knowledge of who owned which animals (whether they were offending against the rules or not) was clearly a key factor in the regulation of upland grazing. It was important in a system of private ownership based on communal grazing management that each farmer should be able to identify his own stock. To deal with this, the fell sheep keeping community developed the *Shepherd's Guide*, a book where the identification marks of all the fell flocks, organised by township, were recorded. A system of shepherds' meets where stray sheep were physically reunited with their owners accompanied it. A great many of the marks that were developed stay in use today. It is interesting also to note that many of the marks and some of the dialect names to describe them are the same in Shetland, Norway and Iceland.[51]

The first *Shepherd's Guide*, covering a part of the eastern Lake District, was produced in 1817, with further ones covering a wider area being published in 1820 and 1829. It is almost certainly the case that not every farm owned one of these books, but rather that parishes or townships owned them for collective reference. The first really comprehensive *Guide* covering the whole of the Lake District (and possibly the first to be owned by significant numbers of fell farmers) was that produced by William Hodgson of Corney in 1849. It contained over 1,100 entries. Hodgson felt that his book would, "prove to be of the utmost benefit to all Sheep-owners who have stocks going upon the Commons." The *Guide* contained rules and implied conventions – that if stray sheep were come across they should be taken care of and that an attempt be made to identify ownership and to inform the owners. The owners, however, had to verify their ownership and this was not (and is still not) verifiable by assertion only. The animal being claimed must have the known and accepted marks of the flock, especially the earmarks.[52]

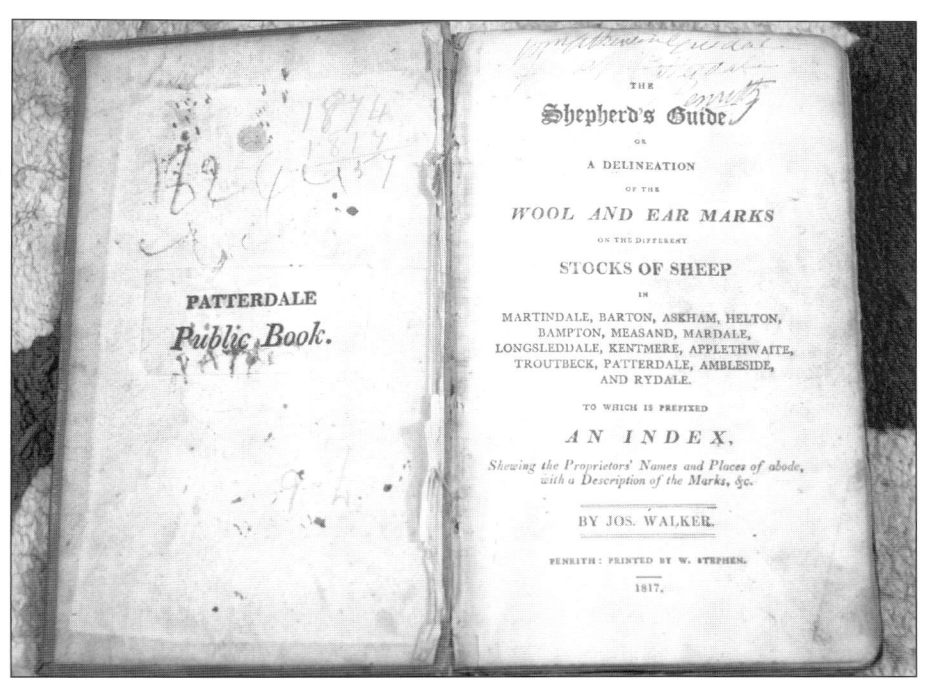

When Daniel Gate of Keswick compiled a new *Guide* in 1879, he wrote in his introduction a few words of explanation for those who were not familiar with a *Shepherd's Guide*. "Owing to the greater portion of our mountains being unenclosed, sheep are daily straying away from their heath, and are often taken up by shepherds many miles away from the residence of their owner; and though there are some hundreds of different marks, the party who has taken up the stray sheep can at once find the rightful owner by referring to the *'Shepherd's Guide'*..."

Virtually all areas of the Lake District where Herdwick sheep are kept have their own Shepherds' Meets. Although they are often only a shadow of their former selves, they keep alive a residual function of the returning of stray sheep alongside an occasion for socialising and the inevitable talk of shepherding. In the 1985 issue of the *Lakeland Shepherd's Guide*, the shepherd's meetings are given as Skiddaw Range, Buttermere, Wasdale, Stoneside, Walna Scar, Troutbeck (Windermere), Mardale, Dockray and Matterdale.[53] In earlier years there was also a Langdale Shepherds' Meet held at the Old Dungeon Ghyll hotel. The meet in July 1921, for instance, was an occasion when 50 'waifs and strays' were brought in from the surrounding fells. R. Birkett and J. F. Buntin Jnr. were in charge of the pens. Once the work of identifying and claiming sheep had been completed there was opportunity for 'jollification.'[54] These meets were once vital institutions, especially in areas where the distance between valleys round by the roads was very much greater than it is as the crow flies.

It was necessary to have such a system which reunited sheep and their owners, especially at the key times of year for the management of stock. The shepherds' meets of the different areas typically have a meet in July and November. The July meet was to ensure that all the sheep were gathered for clipping – in order to harvest the year's crop of wool, a vital part of the economy of the fell farm. The November meet was to gather all the ewes to be put to the tups so that there would be as full a crop of lambs as possible in the spring.[55]

The back-end of the year, therefore, saw a burst of highly collaborative activity: gathering for tup time. Rawnsley, again, provides a well-observed description. He writes about the famous Mardale Shepherds' Meeting, held regular as clockwork on the third Saturday in November at the Dun Bull. He reported that the shepherds "gave up a week to 'raking' the fells and bringing down to the Dun Bull the sheep that were not their own." He reckoned that at that time (the first decade of the twentieth century – just about a 100 years ago) about 200 stray sheep were brought together, "and returned to their rightful masters." The Canon commented to one of the shepherds, "It's good of you to take so much trouble." The reply was, "Naay, naay barn,why it's nowt... ye see it's fair aw roond. They deu t'saame fer me."[56]

In a slightly earlier account, acting effectively as a verbatim scribe into passable dialect, and doing something like oral history, he describes the

Wasdale Head Shepherds' Meet Show, 1960. W. Rawling (far right) one of the judges.

Champion Gimmer Shearling at Stoneside Shepherds' Meet, 1950s. Bred by A. M. Askew, Pike Side, Ulpha, (owner has a cigarette in his mouth). Tyson Hartley is holding the sheep.

shepherds' meet at Stybarrow Head in the late 1890s. He also takes the opportunity to explain the heafing system on the St. John's in the Vale side of the Helvellyn range where there were in fact, "thirteen or fourteen" flocks. Rawnsley describes attending the summer meet on Helvellyn (on the first Monday after the 20 July) and walking up to Stybarrow Dodd above the Sticks Pass for the meeting.[57] The system was described in some detail, once in dialect to Rawnsley (who liked it and transcribed it) and once to a lady visitor for whom the farmer 'switched codes' as the students of linguistics put it, and spoke in Received Pronunciation, the 'Queen's English.' The quotation is a long one but worth reproducing:

"You know the fells on which our flocks are pastured have no walls upon them to separate pasture from pasture: Helvellyn is just one long common, as you may call it, from end to end. The farms in the dales have, by an old kind of prescriptive right, dating from some general agreement centuries ago, so many sheep attached to them. When we let a farm we say it is let with so many sheep, or so much stinted pasture. For all Helvellyn is divided into pastures which are stinted into the number of sheep allowed to graze on them. My stints carry 250 sheep; Bridge-end stints allow of 600; Thirlspot stints carry 250; Dalehead 400 head, and so on. The flocks, by mutual consent, are never allowed to exceed this number, or they would press one on another, and stray off to find pasture beyond their own proper boundary. Indeed, as it is now, if a flock of fellside sheep grows weak and becomes less than its full value, we find that the stronger flocks on either side encroach at once on its pasture. These separate pastures, though they have no walls to divide them, are very clearly marked out by usage and tradition in our minds, and are called 'heafs'; we all know our separate 'heafs', and we train our flocks to know them too."

He also pointed out that sheep could shift from their heafs due to dogs or ponies, and that there was no legal basis to the heafing system: a commoner could graze his sheep wherever he wanted on the common, "but custom and good-fellowship prevent him." Ever the optimist about these matters, Rawnsley quoted a farmer as saying, "our fellside shepherd system is one of goodwill and good neighbourhood, and our meetings help that way."

The same respondent pointed out that it was tempting at times, "to get a nibble off a neighbour's heaf" but generally speaking people resisted temptation. There were, of course, those who infringed the conventions – and several examples were given of sheep stealing cases – but the conclusion was that, "the high fell shepherd has to be a gentleman before he can be a shepherd; conscientious and honourable, and kind and neighbourly. We could never grow wool another year on Helvellyn if it were not so."[58]

Until recently sheep would have to be walked from the shepherds' meets back to their home farms on these fixed occasions every year. With the arrival over the last generation of the motor van and more recently the standard equipment of the pick-up and livestock trailer, the retrieval of sheep

Shepherds' Meet at the Dun Bull, Mardale, 1908

Tups at the Woolpack, 1930s. Photographer Mary Fair - by kind permission of the Cumberland & Westmorland Antiquarian & Archaeological Society and Tullie House, Carlisle.

from distant farms is now a regular and frequent occurrence. The arrival of the telephone on the fell farms has, of course, been the key development here.

It is easy to forget just how recent are good levels of communication and mobility. Isolation was the key factor for hundreds of years – although the railway from the mid-nineteenth century made a great deal of difference in some places. The Post Office also had an impact, with farmers from different dales whose sheep 'met' each other at the fell, often using postcards to send important messages to each other. It is only in the last 40 or 50 years that easy movement of people and stock has been with us. It would have been important that arrangements to return or get back stray sheep were predictable and if possible that they were in a place which was mutually convenient. Given that this would involve considerable distance, and a walk on foot with stock, the opportunity was also taken to combine this with some social activity: the shepherds' meet was born.

This was also the case for other activities which had a basic economic function: the most notable being the walking of tups to the Fell Dales Show at the end of September (and their return in the spring at the May meeting).[59] In the past tups from Buttermere, especially from Gatesgarth, were walked over Scarth Gap into Ennerdale where they joined up with the Ennerdale tups, especially those from Gillerthwaite. They then went over Black Sail and on to Wasdale Head where they met up with the Borrowdale tups (which had come over Sty Head) and the Wasdale tups. Sheep and shepherds stayed the night at Wasdale Head. It was a relatively quick walk the next day over

Burnmoor and down to the Woolpack in Eskdale – where for many years the Fell Dales Show was held – before getting on with the business of letting and hiring tups to keep the bloodlines fresh.[60]

The quest by many breeders for better quality tups is an abiding concern and being able to hire them rather than buy them has been a low cost way over many years of getting access to different genes, thus avoiding in-breeding and it was hoped, getting access to some superior genes which 'clicked' with at least some of your ewes. The majority of tups were hired rather than sold until relatively recently. It is really only in the last 30 or 40 years that tups have been habitually sold rather than let – but some hiring still goes on especially amongst real tup breeding enthusiasts. It has become much more common in the past ten to fifteen years to sell tups at shearling age, so the number that get into the hiring system is nowadays quite limited.

There are, however, some tups which are destined to be sold when they are mature and these might be available for hire in their younger years. Breeders often 'speak for' tups when they are about a year old and they perhaps use them on a few ewes in their first season to see if they are fertile, then 'winter' them well ensuring that they grow sufficiently during the winter and return to their owner in the spring in good and healthy condition. There is usually a bit of renegotiation as to whether the sheep will be let for a further season – which will depend on both the owner and the hirer being satisfied. The size of the hiring fee is not of much consequence with the parties being much more interested in the performance and potential of the animal.

Shepherding
In 1947, William Tyson of Watendlath, then secretary of the Herdwick Sheep Breeders' Association, wrote a small pamphlet about the breed. This work included the obligatory statements on the possible origins of the breed, but he also made a more original and valuable observation:

> "When Herdwicks were first introduced and by whom appears irrelevant compared with the amount of perseverance, patience, and hard work involved in dividing these sheep into separate flocks or stocks and 'heafing' them on those portions of ground belonging to the respective holdings. One cannot help but admire these shepherds of the past and doff one's hat to them for the great task they accomplished in the days gone by."[61]

Obviously, to get sheep to heaf to particular areas would require a great deal of shepherding at the fell. We can get glimpses of what was involved here from various sources over the years – but anyone who has tried to establish or re-establish a flock of sheep on a fell will immediately appreciate the issues. Also, just straightforward shepherding of fell sheep in an age before the arrival of effective chemical treatments (against, for instance, blowfly strike as we saw above) was an arduous enough business. Winchester writes of those many years when sheep flocks were being continually heafed that, "To walk through the hills in the summer months was to pass through a living landscape, in which one might encounter youths tending flocks and herds, poor children gleaning wool, or manorial officers driving stock to the pound."[62] In the second half of the twentieth century all this work in maintaining heafing was factored into the value of surplus stock sheep that were being taken over by a new tenant. William Wilson Jnr. of Bassenthwaite, who did much sheep viewing, claims to have been responsible for introducing the convention of a heafing charge of 50% being added to the bare market value of the sheep.[63]

Even when the tourists first began to appear in the late eighteenth century, a perceptive observer might also notice the continuing presence of people working on the fells. The young William Wilberforce (later the great campaigner against slavery and the slave trade) when on a university vacation tour of the Lake District in 1779, was a case in point, being interested in people as well as in scenery. On 23 September, it being an 'unpromising' day in Borrowdale, his guide ("He was a shepherd") gave him an account of some local matters of interest such as the wadd mine and the destruction of the eagle's nest. He also reported, however, on how sheep were managed on the fells. Wilberforce wrote in his diary, "The sheep will stay about the ground & Hills to which they have been us'd, & they go to fold them once or twice a week."[64]

An extremely well-informed observer of mid-nineteenth century Cumberland, William Dickinson, agricultural writer and improver, who had a small estate at North Mosses, near Lamplugh, was one of many before and since to point out that Herdwick sheep are, "remarkable for attachment to the place where they are suckled" and that once they were heafed this became an important part of the shepherd's ability to manage them. He wrote, "In consequence of this fondness for their heaf or place of breeding they require less of the shepherd's care; and their heaf may be gradually curtailed or extended on any particular side without the trouble and expense of constant herding."[65]

The other side of that coin is that Herdwicks can often exercise their homing instinct when they are sold to new farms and will travel great distances to return to their heaf. R. H. Lamb was of the opinion that no feature was, "more strongly marked in the Herdwick than the homing instinct." Lamb gives several examples of Herdwicks finding their way back to their home fells after having been bought by lowland farmers. He cites occurrences as follows: "from Caldbeck back to Wasdale Head; Kendal back to Watendlath; west Cumberland back to Pooley Bridge; Kentmere back to Seathwaite, Borrowdale,"[66]

The only real guarantee (other than immaculate fences) that Herdwicks will stay put in open fell situations is to buy stock if available from a farm which is heafed on that same fell. There are various examples of this given in the brief flock histories in the 1920 *Flock Book*. For instance, J. Cowx of Town End, Uldale, established his flock in the 1890s by buying ten gimmer lambs from Skiddaw Forest, and some also from the flock of the George and Dragon Inn at Uldale. In the same area the flock of Thomas Teasdale of Chapel House, Uldale, had been founded in 1868 from stock from the Greenrigg, Caldbeck, stock of John Ellwood, with an addition being acquired from Edward Hawell's Longlands, Uldale stock. Joseph Todhunter of Mirehouse also bought stock from Hawell, probably when that family moved to Lonscale. Lord Leconfield established his flock on Skiddaw Forest by purchasing it from Mrs. Cockbain of High Row, Threlkeld. John Cockbain of Causeway Foot, near Keswick, bought his small stock which grazed on Helvellyn from John B. Allison of Low Nest, Keswick.

These examples largely concern Skiddaw and Helvellyn which have generally better and more accessible grazing than that on the western fells. These big northern massifs according to Dickinson produced a type of larger Herdwick. He felt that these sheep often, in fact, received far too much attention in the constant attempt to keep them heafed. He indeed felt that one argument for enclosure of the fells would be that it might reduce the excessive amount of time spent on shepherding by some people. He wrote that, "Some shepherds are at daily pains of taking a few stones of hay on their shoulders, or on horseback, five or six miles to their sheep-heaf, and thus induce the sheep to keep to their heaf in all weathers." He also felt that enclosure might prevent over-stocking and reduce communal conflict. "Much of the jealousy and bickerings, for which shepherds are proverbial, would be avoided, and the continual overlaps of flocks and the houndings and disturbances by the shepherds put an end to."[67]

A similar point was made by H. A. Spedding of Mirehouse in the 1870s when he also complained about the over-stocking of commons and how this, "by causing jealousy among the different flockmasters, leads to much unnecessary dogging and driving, which of course takes a good deal out of the poor sheep."[68]

Sheepdogs and Sheepfolds

Dogs were, and remain, invaluable in much more positive ways. As William Dickinson pointed out in the 1850s, and as every fell shepherd knows, "at all times the dog's services are indispensable." He also pointed out that the bond between a shepherd and a good dog can be very strong, with such dogs being held with great affection.[69] In an extended passage written about the same time he put forward the following testimony which eloquently explains the value of the Cumberland sheep dog:

"With the choice of the best mountain pasture, and the most favourable seasons, the best selected flock would be but a profitless concern without the aid of the shepherd's dog. The whole human agency would be completely baffled to do the work of one shepherd and his well trained dog, among the wily members of a squandered mountain flock. The sheep seem to know, as if by instinct, before they have been many minutes under the charge of a good dog, that all their efforts to break away are fruitless, and they must be driven wherever he is directed to take them. Be they ever so numerous or wild, his sagacity and activity are a full match for them, and his intelligence in comprehending the signs and directions of his master is almost beyond the powers of instinct. The Cumberland sheep dog is not behind his fellows in skill and sagacity, and his value is much greater than is commonly estimated at.'"[70]

There are two key requirements for managing fell sheep: the first being to have dogs that are prepared to co-operate with (and to be commanded by) people in order to gather and herd sheep. The second requirement is somewhere to confine sheep when they have been brought under control. Without sheepdogs, sheep management on the fells would by and large be impossible. Gathering is a key skill and is the fundamental requirement of all sheep management. You must first get hold of your sheep in order to drive them (to move them purposefully in a group) for instance, to clip them, to put them to the tup, to lamb them, to wean their lambs from them, to administer medicinal treatments to them and generally to secure their welfare on an annually repeated and repetitive basis. Gathering is not a species of fell walking: true it involves walking on the fells, but it also may involve some running and probably a lot of standing, and waiting, and looking, and watching (and perhaps even sliding and stumbling and falling).[71]

Although there is a certain amount of work a lone shepherd can do at the fell, it is largely about the activity of a group of people, with dogs, working in a collaborative way moving through a large and uneven piece of ground with individuals operating on different contours and levels of a fellside to drive small groups of sheep to join up with other groups of sheep which eventually 'flock' together. Once this has been achieved, the shepherds get behind them so that they can be driven into an enclosure for sorting and handling. The type of dog used on most fell farms is a reasonably distinct local type, usually smooth-haired and unlike the Border Collie in temperament. The Cumberland Cur is generally much less 'biddable' than the Collie. The Collie tends to be selected for obedience rather than for an instinctive and intelligent ability to seek and gather sheep which is the hallmark of the good fell dog. Cur dogs are often late developers, only becoming 'biddable' at about three years of age.[72]

It is sometimes the case that a cur dog will jump above brackens whilst it is running, looking for sheep, and also that it will work out of sight of its handler sometimes for quite long periods. A useful attribute of a good fell dog, especially in an era of greater bracken growth, is that it barks on command and also when it is in situations where the dog cannot see the sheep and vice versa.

To a great extent the key to effective heafing and the consequent management of the flock at the fell was, until relatively recently, the fellside sheepfold. As Arthur Raistrick put it sheepfolds, "all cater for the same needs – a place to gather, count or check the flock or a place to shelter them."[73] As any fell walker knows, the fells are full of these old structures. Some of them are marked 'sheepfold' on larger-scale Ordnance Survey maps. To fell farmers they also have names, since there will be knowledge that they had a function (probably now largely defunct) for a particular farm and perhaps on a particular heaf. On Kinniside Common in Lankrigg Moss, for instance, there is what is locally known as Richardson Sheepfold marking the time that one branch of the Richardson family worked the township's major farm, Swinside.

On a nearby piece of fell called the Whoap, there is a sheepfold locally known as Laggat sheepfold after the farm which managed its flock in there for many years. They were obviously essential on those very many occasions when it was impractical to take sheep all the way to the farm steading, always more so when the distances involved were large – which they often were. There are many heafs a good five or more miles from home – and in the past some really quite surprising ones. Arising as a result of some complex divisions of the Barony of Kendal dating back to the thirteenth century, it was recorded in Hodgson's *Shepherd's Guide* in 1849 that rights still existed for various farms and stocks of sheep to graze summer fell land resources to the north in the Wythburn Fells, in Scandale, and on Seat Sandal and Helvellyn. Ten stocks of sheep from eight farms south of Hawkshead and between Coniston Water and Windermere lake had rights to graze their sheep in Wythburn in summer – to which area they were accompanied by geese. Another two (Clappersgate and an Ambleside farm) had rights to go in Wythburn Head, and four farms had rights to go in Scandale. George Black from Attwood had a flock that went on Seat Sandal, and George Herdson of Atkinson Ground had one which went on Helvellyn in summer.[74]

By the time of the publication of the 1879 *Guide* only John Croasdell at Skinner How, Hawkshead, and William Hawkrigg from the farm at Sawrey where Beatrix Potter was later to live, were recorded as having sheep which "go in Wythburn in Summer."[75] There is, however, oral evidence (collected from, for instance, Joseph Harrison of Butterilket and Ernest Brownrigg from West Head) by Susan Johnson in the 1960s, that there were still these migrations of sheep (and sometimes of geese too) in the early years of the twentieth century. In 1970 a farmer told Mrs Johnson that he, when young, had annually for nineteen years, driven three stocks of sheep out of Furness, through a piece of Westmorland and then over Dunmail Raise into Cumberland, a distance of about seven or eight miles. He recalled that about a hundred yards over the Raise, "we turned them out through a gate in the east side of the road in Cumberland. Our folds were under Seat Sandal."

Each flock of sheep will have had its sheepfold on their distant heafs.[76]

Not far from Styhead Tarn at the foot of Green Gable on the Borrowdale side, the OS map actually names a sheepfold – 'Patterson's fold'. This is, in fact, a mistranscription by a mapmaker of the dialect rendering of 'Patrickson's fold.' In the mid nineteenth century a Patrickson from a Kinniside farm used the fold as a distant base for managing a flock of sheep. The story goes that the Patricksons made, "a two day journey, spring and autumn, with their stock, through the length of Ennerdale to the part of Sea'waite Fell towards Windy Gap." Joseph Harrison of Butterilket relayed this story to H. H. Symonds in the early 1940s, "Patrickson used to sleep on the fell with his flock when he came to look them over. He got deaf with wet lying and sold his stock to Sea'waite saying: "I don't want others to get deaf sleeping out there.""[77]

Possibly the most famous of all sheepfolds is one of the imagination: the unfinished one by the side of Greenhead Ghyll, in Grasmere, that the old farmer, Michael, began to build after the departure of his son, Luke, for the city. Although Wordsworth invests the sheepfold with great symbolism it also had, were it to be finished, practical use. It was for Michael, "the Fold of which his flock had need." Indeed, while seeking the inspiration for his poem Wordsworth made a number of visits to a Grasmere sheepfold. His sister Dorothy recorded in her journal that on 10 November 1800, "William had been working at the sheep-fold. They were salving sheep." It must not be thought, however, that Wordsworth was helping out at the sheepfold. Judging by other entries of Dorothy's, it is clear that William was 'composing' or seeking the inspiration to do so.[78]

Snows, Diseases and Disasters

Fell shepherding is a demanding business. William Dickinson in 1850 wrote that, "The whole life of the mountain or Cumberland shepherd of modern days consists of extremes of laborious and hurried exertions in times of emergency, and of comparative ease when the flocks are known to be safe."[79] The worst-case scenario was heavy snow. Dickinson recalled the heavy late storm of 1827 that fell as sleet and encased the ground in ice for many days, "till the sheep died of hunger and starvation by hundreds." A flock in Ennerdale lost 300, there were big losses in Buttermere and, "the losses all along the western range of mountains were both numerous and severe."[80]

He also remembered the great Martinmas snowstorm of 1807 which, he felt, was by far the heaviest fall of his lifetime. He was involved in trying to rescue a flock of 400 sheep all of which were buried. A young dog with an aptitude for seeking out the sheep was responsible for the release from the snow of a couple of hundred sheep by 'marking' where they were, but the rest were lost. (This incident reminds us graphically of the importance of the sheep dogs without which fell farming would not be possible, not just in extreme conditions but in the work of every day.) The severity of the 1807 storm can be illustrated best perhaps from the fact that the snow fell on 18 November and that a bull at Thornholme near Calder Bridge was buried in the snow at a depth of 35 feet, "and could not be got out to be skinned before the first of May in the following year."[81] Moving forward 140 years the snow of the winter of 1947 did not fall until early February but it then froze and was topped up by several feet of snow in early March. It took twelve weeks for the road to be dug out to the remote Matterdale farm of Dowthwaite Head.

The fellside farmer, the Herdwick farmer, was – and still is – his own shepherd. Good shepherding involves a constant duty of care, not just in times of bad weather. There is always the next thing to think about: sheep to move, sheep to receive medical treatments, and always sheep to 'look' and their future welfare to mind. There is always the need to ensure, if possible, the availability of adequate grazing resources. Although it is often pointed out that historically the principle of "levancy and couchancy" applied – i.e. that a farm should only carry a level of stock that could be wintered on the farm's own resources, this ceased to apply by the mid-nineteenth century and there began to be a significant connection between fell farmers and lowland farmers especially in relation to winter grazing ('wintering').

Whenever possible all the ewes would be returned to the fell – as Joe Harrison of Butterilket used to say, the fell gate was the most important gate on the farm[82] through which sheep could be sent to enable the meadows and pastures to grow, in the first case for a crop of hay, and in the second case for a good bite to be available when the sheep were gathered in from the fell again. Herdwick farms typically have only small amounts of inbye (fields adjoining the farm steading) compared to their large areas of intake and fell. The inbye land has to generate the feed to sustain the stock during the winter and, therefore, had to be cleared of stock as quickly as possible after lambing time with the geld sheep (sheep without lambs) going to the fell first in the early spring. Amongst these would be the gimmer hoggs (the over-wintered ewe lambs). They will have returned from winter at the end of March or in early April.

The few cows in the byre (and the young stock associated with them), the horses (which took a lot of high quality feeding), the tups and any other sheep (such as wethers and cast ewes) that might be waiting for sale would all need feeding or access to grazing during winter. The policy would be to clear the inbye land of as many animals as possible to accommodate what could not go the fell. The ewes will have been returned to the fell around Christmas after having been tupped. The shearlings ('twinters' – sheep of two winters of age) will have remained on the fell, but since they were considered to be too young to carry and rear a lamb, they would often be 'clouted'. Clouting involves sewing a small piece of cloth (a clout) over the tail area of a sheep in such a way as to prevent it being served by a tup. Needless to say there are only a very few farms where the practice of 'clouting twinters' has been carried out in recent years: the last example I know of being Tilberthwaite under George Birkett's direction.

Securing good wintering for the young replacement female sheep, the gimmer hoggs, has, for instance, always been a vital task and fell farmers often developed long term, annually renewed arrangements with farmers in the adjoining lowlands at a cost currently of about £15 per head for a twenty week period. The gimmer hoggs are the breeding sheep of the future and need to be given good treatment in their first winter, which they usually get in the coastal and lowland areas where they are sent. A key area for hogg wintering for the Langdales, Eskdale, and the Duddon Valley, was the area of enclosed low fells between Coniston Water and Windermere which later became afforested and known as Grisedale Forest. Vic Gregg recalled that hoggs from Low Millbeck, Great Langdale, were walked via Skelwith to Hawkshead or Sawrey (seven or eight miles) and then on the next day to Dale Park and beyond as far as Grizedale or Haverthwaite. Their shearlings went on Holm Fell and Brow Fell for a couple of months. The hoggs were walked back on or just after 5 April – with two or three thousand, from various farms in several dales, assembled between Hawkshead Hill and Force Forge.[83]

Fell farmers from the northern part of Herdwick Country naturally tended to winter sheep on the Cumberland coastal plain or on the Solway, either on the Plain, or on the Marshes. The owners of the grazing rights on the marshes would (and still do) collectively employ a shepherd to look and if necessary move stock in times of high tides and floods. But even here things could conspire against you. On 5 November 1926, Jerry Richardson of Gatesgarth, wrote the following entry in his diary, "Very wild morning. Big flood. Most of snow gone. Killed a sheep. Mending flood work. Sawing wood. Had a wire from Rockcliffe. All Hoggs drowned."

The following day he motored to Rockcliffe, where he found in fact that he had three hoggs left alive out of the 60 of his that were wintering on the salt marshes there. They were not alone but were wintering with hoggs belonging to other people: collectively 1,200 sheep were drowned in this incident. In the winter of 1938, again on the marshes, 1,000 "good, heaf-going hoggs" were caught by a high tide and drowned.[84] The smaller farms also had their share of misfortunes and annual losses. There were always some of the precious gimmer hoggs which did not return from wintering. For instance, John Sawrey who farmed at Grassguards, Ulpha, wrote in his diary in 1895 that his son Edward went to Langdale on 9 March to see the hoggs, "when he got there they told him there was nine dead at John Creaghton and two dead at Mikel Riggs" as well as an unrecorded number at Richard Borrow's.[85] All farms had their losses of this sort. William Dickinson writing in 1850 estimated that the loss by death during the first year of a sheep's life was between five and ten per cent – but that was not the end of it as things could continue to go wrong. "Lucky is he," he wrote, "who can sell, at maturity, 65 sheep from every hundred of his lambs."[86]

Another constant complaint over the years has concerned the depredations of foxes. Lake District fell farmers have often been great enthusiasts for hunting, with six fell packs (until the recent ban) actively hunting the area. Hunting also provided some winter time contact with other farmers at a time of a generally solitary existence. It is certainly a source of great annoyance when foxes take lambs soon after they are born, destroying the year's work, and this can sometimes reach serious levels. Sometimes, probably, foxes got too much of the blame: the late Derwent Tyson recalled to me the situation at Watendlath one year shortly after the end of the Second World War that 900 lambs were marked out to the fell, but only 400 came in. Foxes were part of the problem but it was also due to the ravages of lamb dysentery in an age before vaccination against clostridial diseases was common.[87]

William Dickinson recorded in the mid-nineteenth century that the fox was, "pursued to destruction as the most mischievous of all vermin."[88] Crows were also (and remain) a serious pest, removing the eyes of sheep when they are down, and the tongues, and entrails of new born lambs. The Churchwardens' Accounts for Hawkshead in the time of Wordsworth's schooldays there, reveal that there were rewards for the destruction of vermin including ravens, carrion crows and foxes.[89] Until about the end of the eighteenth century the eagle was also regarded as a pest needing to be controlled, with a rope being kept in Borrowdale on which the shepherds lowered themselves down crags to destroy eagles' eggs. The rope was for the use of the central Herdwick valleys of Buttermere, Ennerdale, Wasdale, Eskdale and Langdale.[90]

One of the great scourges for many years in the Lake District was liver fluke infestation. Liver fluke are particularly prevalent in wet areas and infestation was rife in many parts of Herdwick country. It has historically been responsible for very considerable debilitation of sheep, low performance, and death. An account of farming in Westmorland written in 1909 suggested that the once typical flocks of Herdwicks in the old county were, "fast disappearing, owing in great part to the destructive 'fluke' worm."[91]

The sheep at Troutbeck Park in the early 1920s, to give one example, were riddled with fluke and according to their shepherd, "used to die like flies" but were brought round by the latest treatment (and an effective one): the use of carbon tetrachloride tablets. There has been widespread concern about losses caused by fluke at various times.[92] Herdwick sheep seem to have been particularly prone in the nineteenth century to what contemporaries called 'rot' or 'rottenness'. Describing what seems like a clostridial disease (possibly braxy or more probably louping ill), William Dickinson recorded that, "the mountain breed seems to possess a peculiar aptitude to acquire rottenness by sudden change in the autumn, from a very hard and sound heaf to a new limed or otherwise unsound pasture." There was no cure and Dickinson stated that, "the whole year's breed has been known to be lost from causes of this kind."

Dickinson goes on to list the life of the shepherd in the following memorable terms:

"– of the keenest exposure to drifting and blinding snow and hail, and to the reflected heat of the narrow valleys; – of clambering on hand and knee among the slippery and rugged rocks, and of plunging in the mosses or in the drifted snows; – of boisterous conviviality on a few stated occasions in the year, and of long and solitary wanderings among the mists and rains and perils of the dark mountain tops; – of the wild and incontrollable excitement of the loud echoing tally-ho of a mountain fox chase; – and of the still and lonesome watchings, night after night, near the cubs concealed in the fastnesses of borran and bield. Now toiling alone up the steep mountain side with a large sack of sheep-hay on his head, against a strong and gusty wind; and now singing 'tarry woo', in a full chorus of his brethren, stretched on the sunny green-sward, after the toils of the washing-day are over; – and finally, may be added, after the greatest exertions and most anxious care, obtaining the smallest profits of any class of men of equal capital in the kingdom."[93]

Many of these things will still be recognisable experiences for people who have farmed fell sheep. It must be remembered also that it is only in the past twenty years that there has been a revolution in waterproof clothing and footwear, and that simple things like the cheap Wellington boot have made a huge difference to the sheer level of comfort that people who work outside can now enjoy – to say nothing of the transformation that has been brought about in the past few years by the relative affordability (at least until recently) of supplementary feed and all terrain vehicles. But, on the high fells in Herdwick country, the essential system is one of virtually all year round grazing on the fells with the requirement to bring sheep off the fells for major events and interventions like tup time, lambing time, and for clipping, dipping and weaning. What has also not changed is the nature of Herdwick sheep themselves. They remain active sheep and when sheep are active there must be active shepherds. Samuel Barber writing about the Wythburn area in the 1890s put it like this:

"The 'Herdwicks', as these fell sheep are termed, are a lively race. How comfortably they ensconce themselves among the crags of Wythburn and Borrowdale, cropping the fresh herbage below the jutting crags, and how blithely they skip over the freshly built stone fence, as they contemptuously chuck the rattling stones behind them from off its moss-capped summit..."[94]

He was entirely correct when he concluded that, "the life of a fellside farmer is not *all* poetry."

3: Herdwicks and the Lake District Landscape

"Perhaps more than anywhere else of similar character in Britain the Lakeland fells are used by the general public for walking, climbing and the enjoyment of the open air." L. Dudley Stamp, 'The Lake District' in W. G. Hoskins and L. D. Stamp, *The Common Lands of England and Wales*, (London, 1964)

Part of the Landscape

Herdwick sheep have an almost iconic status in the Lake District. They are an emblem of the landscape, and their existence is almost a proxy for the health of the grazing management system on the fells. Today, indeed, they are seen as a part of the tourism industry's 'offer' to the public. Numerous walkers wax lyrical about seeing Herdwick sheep whilst on the fells and others are interested in seeing the connected farming activity in the valleys. At local agricultural shows throughout the summer Herdwick sheep shows provide a colourful centrepiece – for instance, at Cockermouth, Gosforth, Broughton in Furness, Hawkshead and Keswick, and at the Westmorland County show at Kendal. But shows in the heart of Herdwick country at Ennerdale (last Wednesday in August), Loweswater (early September) and Borrowdale (third Sunday in September) usually have large numbers of high quality sheep from the fell farms.[1] The premier show of Herdwicks, however, is (still) at the Fell Dales Association's Eskdale show on the last Saturday in September.

At the back-end of the year, it is the turn of the shepherds' meet shows particularly at Buttermere, Wasdale Head, Stoneside and Walna Scar, to act as a showcase for the Herdwick – though as tup time gets closer they are mainly events for the Herdwick community itself. The highlight of the Herdwick breeders' year is almost certainly the time of the tup sales, one being held at Cockermouth, and the other at Broughton-in-Furness. These sales, run by the Herdwick Sheep Breeders' Association, take place at the beginning of October each year.[2] About 450 tups change hands annually.

One of the major tasks of the Herdwick Sheep Breeders' Association is to keep up the standard of Herdwick sheep by carrying out a process of inspection and registration of suitable young tups, of which there are each year usually around 350 in number. About 60 farms breed Herdwick tups on an annual basis, with about twenty of these farms dominating the market. A vital part of the Herdwick story is that there should continue to be fell farms where the sheep are kept and farmers who want to keep them. They form a key part of the cultural landscape. There are today still over 100 farms with significant flocks of fell-going Herdwick sheep, but there have been times when there seemed to be great threats to the continued existence of key working Herdwick sheep farms.

Herdwick sheep have been centrally involved in the struggle for the protection and conservation of the Lake District landscape. The continuation of Herdwick sheep farming has been an important part of the arguments which led, firstly, to the creation of the Lake District National Park and, secondly, to the establishment of a permanent reservoir of fell farms where Herdwick sheep breeding has to continue, through the National Trust estate. Three individuals in particular were largely responsible for this: Professor G. M. Trevelyan, Mrs. H. B. Heelis and the Rev. H. H. Symonds. Their relative roles are analysed in what follows.

The Threat of Afforestation

One result of the First World War on the domestic front was the creation in 1919 of the Forestry Commission. This was a state body whose task it was to ensure that in future the country had adequate supplies of timber in case of national emergency. In order to have secure domestic timber supplies, there was, of course, a need to grow trees and to grow trees land was needed – and the land had to be cheap and of relatively low agricultural productivity. In short, land for the planting of softwood trees was to be found in the hills. The fells of Cumberland were one of the prime target areas for the Forestry Commission.

Most of the Herdwick grazing area was, as we have seen, common land, and so could not be purchased for anything other than the continuance of grazing. But there were areas of enclosed fell land that in the agricultural depression of the 1920s became available as a result of sheer financial difficulty on the farms which owned them. By virtue of being made an offer they could not refuse, many acres of enclosed fell were in danger of being planted with trees. Ennerdale, for instance, had been enclosed in the 1870s and the farms all allocated areas of fell which were fenced (using iron posts and wire) against each other and to the parish boundary on the watershed. In addition, Lord Lonsdale, the lord of the manor, received the dalehead ('Ennerdale Dale') and established a Herdwick flock there, building a bothy for his shepherds at Black Sail. In Ennerdale in general, what had been common land was turned into allotments of private property, making it vulnerable to purchase in the time of economic depression that followed less than 50 years after enclosure.[3]

Virtually all the plantable fell land in Ennerdale was thus acquired in the 1920s. The Rawling family of the Hollins was the only family in the dale which did not sell any of their fell at this time. Land was also bought by the

Commission at nearby Blengdale. There was also privately owned fell land available in the Whinlatter area as a result of the earlier enclosure of fell land in Lorton and there was more across Bassenthwaite lake near Mirehouse. In these places trees soon displaced Herdwick sheep. The loss of Herdwick sheep and Herdwick sheep heaf at Ennerdale was by far the most serious with the loss of the Ennerdale Dale stock and most of the very famous Gillerthwaite stock which grazed on the Pillar range and also on the side of the dale which joined Buttermere.

Gillerthwaite was one of the three or four most important sources of Herdwick tups for the whole of Herdwick country.[4] Two thousand Herdwick sheep were lost to afforestation in Ennerdale and about another 1,500 at Whinlatter. In short, Herdwick sheep were having their heafs bought out from underneath them by the State – and on top of this, the heafing system at the head of Ennerdale was also disrupted. As was pointed out at the time, "a whole flock is extinguished and the upper territory is thus left uninhabited, the neighbouring sheep – in this case from Wasdale or Buttermere – will cross the skyline and come trespassing downwards. This blameless and natural escapade leads to great difficulty for their farmer..."[5]

As the inter-war depression moved on, the Forestry Commission acquired more land: notably 7,000 acres in upper Eskdale and in the Duddon Valley with the intention of planting about a third of it and creating the Hardknott Forest Park. This would have included parts of the famous dalehead farms of Black Hall and Brotherilkeld plus a number of smaller Herdwick farms.

All this raised issues of landscape aesthetics and appreciation and loss of public access to the fells – but central to the argument, and put right to the fore, was the argument about Herdwick sheep. The prime mover behind the campaign against further afforestation in the Lake District was the Rev. H. H. Symonds, headmaster of the Liverpool Institute. In April 1933 Symonds had published a guide book called *Walking in the Lake District* and in it articulated strongly the growth of interest in outdoor recreation and public access to the countryside. Symonds was one of the group of highly placed people who began to relate access to the countryside to its preservation and to campaign vigorously for the creation of National Parks.

Symonds had a didactic intention in writing his guide book. It was partly to make the case for National Park status for the Lake District and for the preservation of the countryside, but he wanted to preserve it in, "the best possible way, by teaching as many as we can to use and value it; not by locking it up, or by making a museum of it." He noted with approval the increase in the number of young people in the 'open air' movement and felt that it stood for, amongst other things, good taste, no litter, "for a knowledge of the outdoor world; for an understanding of agriculture and for a decent humility in the presence of the most skilled of all the workers, the agricultural labourer and the shepherd. It stands for the preservation of rural England." In the first edition he had called for people to help set up a body to support this cause.

Tyson and Jonathan Rawling and champion tup, 1930s, at Loweswater Show.

The edition quickly sold out and was reprinted in August 1933. The Friends of the Lake District was formed the following year with Symonds as its leading figure in its formative years.

The Friends effectively used the anti-afforestation case to raise all the arguments which, firstly, led to the agreement extracted from the Forestry Commission in 1936 not to plant in a large central zone of the Lake District and which, secondly, was a key part of the sustained pressure which led to the creation of National Parks in general, and of the Lake District National Park in 1951 in particular. Symonds wrote passionately about, "traditional sheep farming, [and] the Herdwick breed." He made no bones about it, "The Herdwick breed, and those who breed it, must survive." He gave much evidence of the opposition of the dalesmen themselves to the afforestation proposals, quoting at length the correspondence that the Herdwick Sheep Breeders' Association's secretary, R. H. Lamb, had with *The Times*. He noted also that the Annual General Meeting in 1935 had called for, "a protest to be made against the proposed afforestation of good sheep fell in Eskdale and Seathwaite."[6]

For Symonds the first threat from afforestation was the, "fear of still further accumulating damage to the sheep industry"; and secondly, "the fear of damage to the tourist industry," which he felt was the second most important part of the local economy after sheep farming. Although he was enthusiastic about the floristic abundance that occurred in the first few years when, in the fenced off areas, the trees were small, Symonds anticipated that even that

Strays being taken from Gillerthwaite to Black Sail for collection by the Naylors of Middle Row, Wasdale Head. Note Ennerdale before the trees.

would disappear "under the shadow of a wood."[7] Afforestation had a negative impact both on the way an area looked and also on access to it. "It is obvious," he wrote, "that if the Lake District becomes covered with Ennerdales, it ceases to be what it has so far been – the best of the national playgrounds."[8]

Others followed Symonds' line of argument. For instance, in a pamphlet revealingly called *Lakeland, a playground for Britain*, published by the Fell and Rock Climbing Club of the English Lake District in the early 1930s, a case was made for the Lake District becoming, "a great National Reserve" partly on the ground that, "complete freedom for the fell-walker did not in the least interfere" with sheep farming which was, other than catering for tourists, the "main occupation of the natives."[9]

National Property, National Trust and National Park

Today, the National Trust owns about one quarter of the Lake District, including over 90 fell farms about half of which are regarded, to some extent at least, as Herdwick sheep farms. Initially small parcels of attractive and strategic land were bought, but from the late 1920s working, traditional fell farms began to be acquired. A key driving force behind this development was G. M. Trevelyan, Regius Professor of Modern History at Cambridge University, and the most popularly appreciated social historian of his day. Trevelyan was a great enthusiast for the English Lake District, developing a deep attachment to the area, noting in his *English Social History* that it was in the eighteenth century that, "the beauty of Wordsworth's homeland attained the moment of rightful balance between nature and man."[10] Trevelyan was one of many in the academic and political worlds who fully signed up to Wordsworth's famous vision in his *Guide to the Lakes* that the Lake District might be, "a sort of national property, in which every man has a right and interest who has an eye to perceive and a heart to enjoy."

Trevelyan had a strong personal affection for the area. He had been a member of many vacation reading parties based in the Lake District and had bought Robin Ghyll in Great Langdale as a holiday home for his family before the First World War.[11] In 1929 he bought and gave to the National Trust the Dungeon Ghyll Hotel (which was a farm as well as an hotel), and Stool End and Wall End farms. In 1938 the Trust itself bought Middlefell and in 1944, Trevelyan also donated Harry Place and Mill Beck.[12] Others followed suit. In the Duddon Valley, Sir E. D. and Lady Simon, donated Cockley Beck, and Sir Noton and Lady Barclay donated Dale End in 1929.

A local person who anticipated developments of this sort was the Whitehaven businessman and tannery owner Herbert W. Walker who, in the years immediately after the First World War, began to acquire strategic farms and associated fell land where afforestation might have been a threat. Walker acquired land at Wasdale Head (including Burnthwaite) and at the head of Borrowdale (including Seathwaite). Although Walker did not pass land to the Trust (with the exception of the small Rampsholme Island on Derwentwater) it often found its way there over the years. He sold Great Gable, Great End and other fell land at a nominal sum to the Fell and Rock Climbing Club which in turn, in 1923, gifted the property to the National Trust.[13]

It was Trevelyan, however, who most significantly encouraged others to give land to the Trust. It was he who effectively triggered a process by which the larger part of Great and Little Langdale became the property of the Trust with the acquisition of High Birk How in 1948, and Fell Foot in 1957, in addition to the farms mentioned earlier. Side Farm was given in Trevelyan's memory in 1963, and Robinson Place was bought by the National Trust itself in 1974. As a result Herdwick sheep keeping still dominates the agriculture of the Langdales.

From 1928 until 1949 Trevelyan was at the centre of things in relation to the acquisition nationally of land for the National Trust (for instance in the Peak District and on the South Downs) in his role as chairman of the National Trust Estates Committee. It was in the Lake District, however, that the most significant and sizeable acquisitions were made where the National Trust acquired over 10,000 acres. Piecemeal land acquisition of a protective character was only part of his agenda. He also argued very forcefully that in order to preserve the beauty of rural England there needed to be, in his words, "a State policy, the support of the Ministry, of Parliament, and of legislation."[14]

Access to dramatic and challenging countryside was, for Trevelyan and many others of his socially responsible make-up, an important part of the available compensation for the negative economic, environmental and social consequences of industrialisation and urbanisation. Speaking at the CPRE national conference in 1937 Trevelyan put access to countryside very high up the national debate when he stated that, "the condition of any real value in modern city life is holidays spent in the countryside." In terms that today seem almost unbelievable, he went on to say that with shorter hours, paid holidays and increasing leisure for millions, "the question of the proper use of leisure has become a national problem second to none in importance" – with the creation of national parks being "increasingly and urgently necessary."[15]

The people of the world's first industrial nation – with its rapidly expanded towns, squalid conditions, mean streets, poor housing and hard labour – deserved to be able to ramble in the countryside breathing clean air whilst re-creating themselves. Many a dweller in the great northern towns discovered for themselves the beauty of the English Lake District and other areas. The young Alfred Wainwright, for instance, when he was taken by an uncle in the early 1930s from Blackburn to visit Windermere, had his eyes opened in a way that was to change his life – and later to change the lives of thousands of others as his guide books to the fells gave people the confidence to tackle them.[16]

Trevelyan in his autobiography of 1949 wrote of these years that, "The taste for mountain scenery grew *parri passu* with the Industrial Revolution." There had been numerous unsuccessful attempts to get a satisfactory Access to Mountains and Moorlands Bill through Parliament. Individuals such as T. Arthur Leonard, a Congregational Minister, from Colne in Lancashire, from the 1890s began to introduce young working class people to the delights of outdoor recreation. He formed the Co-operative Holidays Association and then in 1913, the Holiday Fellowship. One of the CHA's first centres was at Newlands and many other guest houses throughout the UK were subsequently developed.[17] Popular demand for open-air activity in the countryside was strong. Rambling and hiking, youth hostelling and camping all grew strongly in the 1930s as mass unemployment scarred the economic and social fabric of the manufacturing towns of the north of England.

Trevelyan's belief in the beneficial effects of access to, and fairly Spartan physical exertion in, the countryside led to his being President (between 1930 and 1950) of the Youth Hostels Association, a movement which enabled people on low incomes from the great cities to visit the countryside. He put huge emphasis on the importance of, "strenuous open-air holidays" and asserted that, "nothing was of greater importance to our civilisation than the love of nature, and the ability to indulge and develop it on holidays."[18] The YHA grew very fast from its inception in 1930, establishing 297 hostels (many of them in the Lake District) for the use of its membership of over 80,000 people by 1939.[19]

There was a negative side of things, though not one produced by rambling

The Birkett family leaving Gillerthwaite, loading tups for sale, 1940.

and youth hostelling. The relative isolation of the Lake District began to diminish as better off people acquired motor cars and this, in turn, seems to have stimulated the demand for the building of lakeside villas and other inappropriate developments. Trevelyan felt that the motor car had the negative potential, in his view, "to turn every beauty spot into an eligible building site." Expressing the same impulse as had motivated the formation of the Council for the Preservation of Rural England in 1926, he argued that protection of 'amenity' was vital and he pressed for legislation on national parks, not on the American model but on a specific English (and Welsh) journey which fully endorsed the role of traditional farming both in relation to landscape protection and to recreation:

> "If the Lake District, for instance, were to be turned into a National Park, ownership would be undisturbed, and agriculture and sheep-farming would continue as it does now. Indeed, the farms are part of the beauty of the landscape."[20]

Views of this sort became central to the whole movement for national parks and also to the growing profession of town and country planning. One of the founding fathers of town and country planning, Patrick Abercrombie (along with a colleague Sydney A. Kelly), produced in 1932 a major statement of what this might mean for Cumbria through the publication of a large volume, the *Cumbrian Regional Planning Scheme*. This scheme had effectively been commissioned by all the local authorities (at county, city, town, and Urban and Rural District level) in Cumberland and also had the involvement of

Westmorland County Council. In addition it had the participation of other bodies such as the Lakes Artists' Society, the Cumberland and Westmorland Antiquarian and Archaeological Society, the English Lake District Association and the Lake District Safeguarding Society.

A conference had taken place in Cockermouth in 1929 which commissioned the work and which had Ewart J. L. James as secretary. James had been the author the previous year of a booklet which had highlighted the dangers of ribbon development and had advocated regional planning as the antidote to this and the scourges of filling stations and advertisements.[21] The concerns of the local authorities were clearly how much-needed development might be made to fit alongside the aesthetic considerations that the new profession of town and country planning espoused.

The Regional Planning Scheme basically advocated zoning – i.e. that development and exploitation of the region's natural resources should be allowed to happen where it was not damaging to the landscape. There are sections on all the sub-areas of 'Cumbria' (it took until 1974 for the new county of Cumbria to be created) and they are largely uncontroversial and indeed consist of many statements of the obvious. When it came to west Cumberland, where the problems of the inter-war depression were most serious, and the situation seemingly almost hopeless as far as employment went, the study reported that in 1921 there were about 3,500 men in employment in mining, quarrying and metal working in the Egremont, Cleator Moor and Frizington areas. This was in stark contrast just over ten years later when there were only about 450 men in employment in the same industries. The report commented that there was neither prospect of new industries being attracted to these areas nor any hope of a revival of existing industries.[22]

But the report had a great deal to say about Lakeland. In fact it is its most serious concern, which might have something to do with the fact that Kenneth Spence took over as secretary in April 1930 after James died. Part Two of the study is headed 'Detailed Considerations. The Lakeland Area' and it begins with a detailed development of the case for Lakeland being "A National Property." The study is also highly supportive of traditional farming – in a section on 'Farming and Local Life' it speaks of "native local life and occupation" as a continuing thing with "its interests, its sports, its gatherings..." and of a need to ensure that "nothing is done to interfere with the convenience of farming." It also goes on to commend Trevelyan for presenting some farms in Langdale to the National Trust, "that they may be safeguarded in their present use."[23]

However it also puts a gloss on newer, contemporary problems and concerns from a highly mounted aesthetic position. It draws distinctions between (good) ramblers and (bad) tourists. It complains about the problems caused by mining and quarrying, the extraction of water, the creation of new roads, the erection of further inappropriate (essentially non-vernacular) buildings and also of 'disfigurements' of various sorts. For example, at

Above, Borrowdale Show, and, below, Broughton Tup Sale Show, 2006; photographs courtesy of Dorothy Wilkinson.

ISAAC BENN, Simonkell; forked near, a pop on near shoulder, and one on tail head.

Upper fold bitted both ears, a pop behind the head, and a short stroke down the far side.

Above, entries from Gate's Shepherd's Guide of 1879 and below, two fine young tups at Tilberthwaite in their show red; photograph courtesy of Dorothy Wilkinson.

Above, Yewdale sheep waiting to cross beck after gathering; photograph courtesy of Dorothy Wilkinson.

Below, Walna Scar Shepherds' Meet, Sheepdog Show, 1979; photograph courtesy of Bruce Wilson.

Above, a ewe teaching her lamb the traditional heaf; photograph courtesy of Ian Brodie.

Below, Eskdale May Meeting, 1979. Joe Cowman judging with, left to right, Harry Robinson, Tyson Hartley, William Bland, Gowan Grave, Syd Hardisty and Dorothy Sharp. Thomas Richardson and Willy Greenup in conversation; photograph courtesy of Bruce Wilson.

Lorton there was a "garish wooden shop and some other disfigurements." The arrival of the motor car seems to have created many of the concerns, not only because cars broke down isolation, but also because they needed facilities, which brought the need for signs. A typical complaint was one about Glenridding where there were, "two striking petrol advertisements" and another about Patterdale where the petrol pumps had "entrails laid bare as for an operation."[24]

The central positive proposition was the establishment of a national park for a) preservation of the scenery, b) maintenance of flora and fauna and c) provision of recreation. The maintenance of traditional farming use was effectively the backdrop to these things and the central issue of this part of the study is about how to achieve them. In a section headed 'Methods of Obtaining the Desired Results', the authors go far beyond mere advocacy of an improved planning system to state unambiguously that, "The most complete method for the purpose would of course be the purchase of private property."[25] So, ideally, public ownership would be the preferred option but this was immediately discounted on the grounds that, "the expense involved even for wild and barren land would be prohibitive." They went on to proffer a more realistic option – namely the further acquisition by and donation of land to the National Trust (for which the Lake District was 'the cradle' of its activities).[26] They hoped that, "these ownerships will continue and will include both agricultural as well as wild country, as is the case in the Langdale valley. There are certain places which everyone must feel should be in full public ownership."[27]

The Lake District was rather a special case in relation to the agitation for the creation of national parks. Public access to the fells was very largely uncontested and raised none of the issues that popular struggles over access to the countryside raised in the Peak District – an area that, of course, is much closer to large urban populations than is the Lake District. The fights over access in the Peak District had much to do with grouse moor management. Kenneth Spence, writing in the very influential work on countryside protection, *Britain and the Beast*, wrote of, "the happy absence of grouse or other game" as being a key to the allowing of *de facto* public access to the Lake District fells. In his contribution on 'The Lakes' he wrote of this "unique area of England" which exhibited "some very special problems in rural preservation" and also that it was, "the birthplace of such problems inasmuch as it inspired Wordsworth's *Guide to the Lakes* which should still be bible to all who care for the aesthetics of the countryside."[28]

Spence wished to see strong controls over the use of land brought about through 'a National Park scheme', in the Lake District amongst other places. He felt that the National Trust's practice of "sporadic ownership, which can never hope to be brought to completion," was not only inadequate for the task in hand but was also only achieved at "exorbitant cost." This approach, he went on, even had the serious danger of being a "palliative, or even a narcotic" which would "tend to lull us into a false sense of security." Much more likely to be effective and capable of being rolled out on a large scale was, he thought, the use of restrictive covenants as had occurred at Buttermere.[29] When in 1934 the Marshall family's extensive estate in Buttermere came up for sale, a scheme was worked out between the National Trust, Lake District Farm Estates (of which more shortly), Balliol College, Professor A. C. Pigou, Trevelyan and others, whereby the Trust was to raise as much money as possible to buy as much land as possible, with the others agreeing to buy the rest and subjecting themselves willingly to restrictive covenants. Between 1935 and 1937 covenants were given over in excess of 4,000 acres in the Buttermere-Crummock-Loweswater area including the great and ancient dalehead farm, Gatesgarth.[30]

Although the covenants method was much praised in the 1930s, in reality it was to turn out to be quite limited in its beneficial effect. Covenants may have stopped inappropriate development but they did little or nothing to protect and sustain traditional farming. Covenants have to be negative in substance, protecting land against, for instance, development or afforestation. As Elizabeth Battrick explains, "a positive covenant would not run with the land so as to be binding on successive owners." It is not possible to say that such and such a farm should be maintained as a farm – which, it seems to me, is precisely what is needed to protect the world of Herdwick sheep breeding and the cultural landscape. This committed landownership is something that, of course, the National Trust estate, by and large, delivers.

It was this bundle of concerns – the ownership of fell farms to control land and to retain it in traditional farming use; the extension of public access; the effects of the car on the countryside – that led to the formation of Friends of the Lake District in 1934, and also eventually to the creation of the Lake District National Park in 1951. Some people, however, felt in the 1930s and 1940s that the formation of a National Park would obviate the need, not only for covenants, but also for additional land ownership. B. L. Thompson, for instance, felt in 1946 that if the National Park was created, "the National Trust's need to extend its properties would presumably cease within the National Park area because all problems of preservation and access would be settled by the National Park Commissioners."[31]

This now seems like a thoroughly naïve view, but it was characteristic of its time. The continuing ownership of the fell farms by families and institutions committed to hill farming, which is the key to the management of the cultural landscape, was regarded as largely unproblematic if there was National Park status. There was a widespread view that National Parks would obviate the need to protect traditional farming through the purchase of farms. But, as Ann and Malcolm MacEwen put it, the National Parks Act of 1949 was based on some mistaken assumptions. These had been spelt out in the Scott Report on rural land utilisation of 1942 and further accepted by the Dower and Hobhouse reports of 1945. One of the assumptions was that development control could deal with most of the important issues and another main assumption was that farming and forestry need not be part of the development control process, "because farming would preserve both the rural landscape and community." The Scott report had declared that, "there is no antagonism between use and beauty."

In a book of 1959 celebrating the tenth anniversary of the National Parks and Access to the Countryside Act, Ulverston-born the Rt. Hon. Norman Birkett (another of the small group of influential advocates on behalf of outdoor recreation) was still able confidently to assert that the creation of the Lake District National Park had, "made no difference to the settled accustomed way of life that has gone on for generations, and no change in the ownership of land." Farming (along with tourism) was the key activity of the Lake District and he added that, "the mountains of the Lake District area are grazed by Herdwick sheep, famous for their wool and mutton, and for the way they triumph over the severe conditions to be encountered on the high fells."

It was what would be called today a 'win-win' situation. The beauty of the area was the joint work of Nature and of Man. Nature had made the shape of Helvellyn, Pillar Rock and the Langdale Pikes, but much of the beauty was man-made. The land had been farmed for generations and this was responsible for the drystone walls and the patchwork of "ploughlands and pastures." He asserted that the coming of the National Park had done nothing to alter the familiar scene, but rather would "do much to preserve it for future generations and to make beauty and utility walk hand in hand."[32]

Birkett, Symonds, Dower, Trevelyan and others at the centre of the agitation for National Parks did not anticipate the problems that would arise from pressures on farms to intensify in an era where the imperative of production was paramount; and nor did they seem to anticipate the long term economic vulnerability of the hill farming sector.[33] A core assumption was the unproblematic continuation of hill farming. National Park protection did not, however, confer everything that its advocates hoped.

The Tale of Mrs. Heelis

William Heelis, solicitor of Hawkshead, died in 1945 having outlived his wife by just over a year. In the will of his wife, Helen Beatrix Heelis, following her death in December 1943, he had received a huge bequest of fifteen farms and over 4,000 acres for the period of his life, after which it was to go to the National Trust. He decided, however, to forego this life-time legacy and donated it instead straightaway to the National Trust. This large amount of land had been gathered together (with his considerable assistance and advice) by his late wife over a lengthy period. The will also famously stipulated that the sheep on her fell farms, where the flocks were fell-going, should "continue to be of the pure Herdwick breed." Although the overall Heelis bequest was a large one, it must be pointed out that the Herdwick legacy and ongoing commitment only really applies (by my reckoning) to the three farms of Troutbeck Park, High and Low Yewdale and Penny Hill in Eskdale – for these were the only farms with flocks of fell-going Herdwick sheep in the Heelis legacy.[34]

Known always to fellow Herdwick sheep breeders as Mrs Heelis, she is, of course, much better known to the rest of the world as Beatrix Potter, the writer of children's stories. She had started to develop a concern about the loss of fell farms and of the traditional Herdwick sheep-keeping systems in the early 1920s, acquiring at this time one of the largest fell and Herdwick farms in the Lake District, Troutbeck Park. She was also the tenant at Tilberthwaite near Coniston, where she had a manager. She began through the work of her shepherds, especially Tommy Storey who after a short spell at Troutbeck Park looked after her farm at Sawrey, to breed Herdwick sheep of some quality. During the 1930s her sheep were shown at all the important Lake District Herdwick shows, where the female sheep achieved outstanding success.[35]

There is little doubt that Mrs Heelis became an astute and knowledgeable land manager who had a deep understanding of the issues. She totally understood the importance of the heafing system. For instance, at Tilberthwaite, Coniston, the outgoing tenant had refused to sell her the heafed flock when she took over the farm. However, she decided to re-establish it and encouraged her shepherd Tommy Stoddart to work to do so over the years. When the Stoddarts went into Tilberthwaite in 1934, Mrs Heelis said that although she did not expect a man to do the impossible, "If you can get the hoggs heafed on the sunny side of Wetherlam, without heavy loss, I shall be *well satisfied*."[36]

The Stoddarts succeeded in the basic task quite quickly. When Mrs Heelis was in hospital facing a serious operation in March 1939, she wrote to the National Trust (which was her landlord for the farm) giving notice that she wished to relinquish her tenancy. She wrote, "If & when I retire the Trust should purchase a sufficient landlord's stock of sheep – it would be wicked to let them be dispersed a second time after the labour and profitless expense incurred by the shepherd and me, in founding a new heafed flock."[37]

When she took over the very large Herdwick fell farm, Troutbeck Park in the early 1920s, where the flock was neglected and the farm run down, she allowed her shepherds to build up both the quality and the quantity (700 or 800 draft ewes would be sold from the farm annually) of sheep. When she began discussing the terms on which she would leave Troutbeck Park to the National Trust, she wrote to the Trust's secretary, "At present I am minded to fix the number of "landlord's stock sheep" at the number of 1,100, viz 700 ewes, 180 twinters, 220 hoggs, all to be pure bred heafed Herdwicks… there must always be a large landlord flock at the Park because the tops are not fenced. If a tenant got into debt and the sheep flock were depleted, the heaf would be stolen and lost."[38]

Mrs Heelis and Tommy Storey with 'Water lily', one of her prize-winning ewes.

She had also seen the problems that the depressed economic climate of the times brought with it. There was no money in fell farming, with even mere survival being difficult. The smaller farmers were often forced to asset strip their farms, leaving them little with which to make money and making them vulnerable especially to being bought out by non-farming interests. She wrote, for instance, to Canon Rawnsley's widow, Eleanor, in 1934:

"I take such a pessimistic view of our local farming – *unless under a good landlord*, that I wish there may be a sufficient representative number of the old farms in the hands of the Trust: and no doubt Balliol College will also preserve them. Manchester Corporation & the Forestry Commission are protective against *building*, but they destroy rather than preserve the character of the countryside. And smallholders are hopeless. First they sell off the sheep stocks; and then they cut all the scrub timber, & concentrate on hens."[39]

Fully realising the difficulties of livings being made, she encouraged what today would be called 'diversification' and at Yew Tree, Coniston, for instance, her tenants opened a tearoom. She noted that, "the farm can scarcely pay without the teas & visitors." Many a well-located fell farm began in these years to cater for tourists the income from visitors was, no doubt, doubly welcome because it was 'cash in hand.'

She was familiar with Herdwick sheep even from her early days of lower ground farming at Hill Top, Sawrey. "They are beggars to ramble, these hill sheep. I got two back from Coniston last winter that were making tracks for Scawfell where they were born," she wrote in 1918.[40] Having a large fell flock (which by virtue of sheer numbers was likely to contain some good Herdwick sheep) and a good low ground farm was, in many ways, a perfect combination. She explained to an American publisher how she worked Troutbeck Park in with Hill Top. She had two sheep stocks, "the old pedigree flock" being at Sawrey with the "main flock" on the fells."[41] She in fact had three farms devoted to the production of Herdwick sheep, the third (as we saw above) being at Tilberthwaite which she reinforced with sheep from Troutbeck Park as she enabled her manager there to rebuild its heafed flock. Lamb's *Shepherd's Guide* of 1937 contains within their respective parishes the following entries, "Mrs Heelis, Troutbeck Park"; "Mrs H. B. Heelis, Tilberthwaite," and "Helen Beatrix Heelis, Hill Top, Sawrey."[42]

None of the female sheep that had been initially assembled at Sawrey when Tommy Storey went there to manage the farm were good enough to show – a point that Storey made right from the outset, even though this was a judgement which annoyed Mrs Heelis very much. But Storey was adamant and, after a while, was allowed to do things his way. Some good quality

Saddleback Wedgewood, owned and bred by J. Cockbain, High Row, Threlkeld. Winner of 43 First and Champion prizes from 1924 to 1928.

stock was bought from the Greggs (who were on the verge of leaving Town End at Troutbeck where Tommy Storey had worked before he went to Troutbeck Park) and some surplus stock from Troutbeck Park itself. She also asked Joe Gregg to buy her the best tup he could find and, according to Vic Gregg of Great Langdale, he bought her one from the Hollins at Ennerdale.[43] Later she was able to hire such exceptional tups as Jerry Richardson's Gable Blueboy and Joe Cockbain's Blencathra Wedgewood, "perhaps the greatest Herdwick ram of his day."[44]

Mrs Heelis joined the Herdwick Sheep Breeders' Association in 1925. In Volume 5-6, for 1924-25, of the Flock Book the list of flock owners includes, "No. 220, Heelis, Mrs. Wm, Hill Top Farm, Sawrey, near Ambleside. Prefix Claife."

She attended a number of Association meetings in the 1930s. The first meeting she attended was an Extra-Ordinary General Meeting held at Mitchell's Auction at Cockermouth on 2 March 1931 where a ewe inspection scheme was agreed to. This, however, was a very thorny issue and necessitated a further meeting – which she also attended - on 11 May, before a scheme was established.[45] She usually had Tommy Storey with her in attendance. This was the case, for instance, at the Annual Meeting at the Queen's Hotel, Ambleside, in March 1933, and the following year at the George Hotel in Keswick. She did not attend the 1935 AGM, although Storey did. Neither of them was present at the annual meetings in 1936 or in 1937. She was present in 1938 at the Queen's Hotel, Ambleside, where there was a lengthy discussion about the "maggot fly pest."[46] She did not attend the 1939 Annual Meeting – she was frequently ill during this period – and at a Council meeting a couple of months later, it was reported by the secretary that neither Mrs Heelis nor Mr Heelis could accept office as president-elect. Dick Wilson of Glencoyne was chosen instead.

She was not at the 1940 AGM where there was a strong protest about the price being given for Herdwick wool and a debate deploring the call-up of young fell shepherds for military service. Nor was she present in 1941 or in 1942. She did, however, attend the AGM at the White Lion Hotel, Ambleside, in March 1943. There were problems about the availability of fuel at the time and very few were present. The record of the meeting recalls that, "in the absence of both the president and the president elect, Mrs Heelis was moved to the chair." Shortly afterwards in the meeting she herself was selected as President Elect: that is to say she would become the Association's President in 1944. She explained the circumstances to Joseph Moscrop, "There were only six present owing to petrol, they put me in the chair; I did *not* drink porter at the White Lion!"[47]

With this incipient office she was entitled to attend council meetings of the Association and, therefore, was at the Bank Tavern on 20 May 1943. In the absence of other office holders she took the chair. The meeting consisted of a debate about the new hill subsidy payments which the association wished to see being extended to gimmer hoggs (ewe lambs). Those attending the meeting were R. M. Wilson, William Wilson, S. Brownrigg, J. Teasdale, J. Richardson, T. Bainbridge, W. Rawling and R. H. Lamb. She did not, however, attend the council meeting of August 1943. The next one took place at the Bank Tavern, Keswick, on 15 January, 1944. William Wilson was in the chair, the meeting having been called on account of the death of R. H. Lamb, the secretary, who had died of a fractured skull at the age of 52 after being knocked down by a bicycle in Caldbeck.

Wilson in his remarks, "also referred to the death of the President Elect Mrs. H. B. Heelis." There was a vote of condolence to the families of both the deceased office holders, with the members present standing silently as a mark of respect. It was agreed that a letter of sympathy be sent to Mr Heelis and the Misses Lamb. William Wilson, once again, stepped into the breach and became secretary until such time as a new one could be appointed. If she had lived Mrs Heelis would have become President at the AGM in March 1944. Instead, at that gathering, Jos. Teasdale of Caldbeck was elected President, and William Tyson of Watendlath was installed as secretary. Mrs Heelis's death was reported alongside not only that of Lamb, but also that of other respected flockmasters, John Tyson of Ennerdale, John Tyson of Ulpha, and William Riley of Bootle.

The Association Flock Books record the tups that were registered from her farm. There were two born in 1932, two in 1933, two in 1934, and six in 1935. The tups born in 1935 were all sired by Jerry Richardson's tup, Gable Blueboy which won the Keswick May Fair three times in succession. This tup must have 'clicked' with the small number of ewes which she bought from Wilson Allonby of Howe Head, Coniston, since five of the six were out of them. In spite of the fame of her own female lines, they do not seem to have been great breeders of tups, though she did win the prestigious Edmondson Cup for the best tup shearling at the Keswick May Fair in 1934.[48] Generally, however, Claife tups were neither numerous nor particularly celebrated in the Herdwick world, though there were nine tup shearlings got at Gatesgarth in 1946 which were sired by a Mrs Heelis tup. There were a few more tups bred in the Claife flock in the late 1930s: one in 1938; two in 1939 and three in 1940. The last registration of a Heelis tup is that in 1946 for Claife Billy (sire, a Scowgill ram from Darling How, Lorton) born in 1943 with the owner being recorded as the late Wm. Heelis, Hill Top, Sawrey.

Mrs. Heelis's role in connection with Herdwick sheep should be kept in proportion. She certainly did not save the breed from extinction: it was by no means close to this during her lifetime; she was not the most prominent breeder of her day even though she had success in the show ring; and her involvement with the Herdwick Sheep Breeders' Association whilst significant, was relatively slight. Her life, however, from the mid 1920s was about Herdwick sheep farming and the ownership and management of Herdwick sheep farms. She was thoroughly immersed in the Herdwick culture and appreciated within it. She fitted in with the community, but it certainly did not revolve around her. She was not regarded locally as a sort of latter day saint that she often seems to be thought of now. Perhaps the best and simplest way to put it is as Betty Birkett put it to me in 2002 when she said, "she liked being amongst Herdwick men and they liked her." She was a definite and respected part of the Herdwick sheep breeding community in the 1930s and early 1940s.

The leading lights within the Herdwick community were such people as Betty's father, Jerry Richardson, along with Dick Wilson, William Wilson, Jos. Teasdale, Thomas Bainbridge, old Edward Nelson, Isaac Thompson, Jonathan Rawling, William Rawling, Thomas Bowes – these and others were the elder statesmen. Mrs Heelis was accepted amongst them. She clearly learnt from them and took their advice, for instance on who should get certain tenancies, 'head-hunting' a number of good shepherds to important farms.[49] It is clear also that she understood the significance of what they did: she is on record in fact as having felt that she was the only one who understood it. She explained to an American about her semi-autobiographical book *The Fairy Caravan* in the late 1920s that, "I am conceited enough to say I am the only person who could have written about the sheep; because I know them and the fell like a shepherd; but the Herdwick men are not articulate."[50] She sometimes also reacted strongly to their failure on occasion to agree with her. She wrote in 1930 to Joseph Moscrop, a shepherd from north Cumberland who came to Troutbeck Park every lambing time, that, "I think these Herdwick shepherds are annoying men." On another occasion she wrote to S. H. Hamer, the secretary of the National Trust, complaining that "Herdwick men are untidy farmers."[51]

Be it as it may that she felt that Herdwick men were annoying, untidy or inarticulate, she had the foresight (with others) to try to protect the integrity of what might be called 'Herdwick Country.' She was one of the handful of people who, having realised that the resources of the fell farms and the Herdwick flocks were threatened by the economic circumstances of the times, took action to protect them. She bought the large Monk Coniston estate – some of which she immediately sold to the National Trust and which she managed on its behalf and some of which she kept for herself and which went to the Trust after her death and that of her husband. She also bought farms in Eskdale, Hawkshead, and Little Langdale.[52] Her husband only survived her by a few months and it was not long before the landholding that the couple acquired between them – a significant group of farms and about 4,000 acres – passed in the early 1940s to the National Trust. The majority of the farms were small low ground farms in the Hawkshead area.

The legacy of true fell farms with flocks of fell-going sheep is, as mentioned earlier, surprisingly small. Although her will stated that, "the landlord's sheep stocks on my Fell Farms shall continue to be maintained of the pure Herdwick breed," this concerned surprisingly few of her farms. It concerned only Troutbeck Park, High (and Low) Yewdale in Coniston, and Penny Hill in Eskdale, though it could justifiably be added that Tilberthwaite should be added to the list. Mrs. Heelis had bought it initially as part of the Monk Coniston purchase before selling it to the National Trust and it was in this sense also part of her overall fell sheep farming legacy. She had, after all, been its tenant and had ensured the rebuilding of the heafed fell flock and its proper management.[53]

Lake District Farm Estates

The Rev. H. H. Symonds, the prime mover behind the Friends of the Lake District, personally acquired what he regarded as sensitively-located fell farms especially in the Duddon Valley during the 1930s. He put his own money where his mouth was, as he continued to maintain pressure on the Forestry Commission not to plant trees on land in the western dales which was excluded from the 1936 agreement. He campaigned particularly vigorously about the Butterilket land in Eskdale brokering a covenant in 1943 between the Forestry Commission and the National Trust that it should not be planted.[54]

In 1950 Symonds gifted to the National Trust a group of five farms in the Duddon Valley and Ulpha. Thrang, Browside, Hazel Head, Brighouse, Pike Side and Beckstones made up in all nearly 600 acres of in-bye with an additional 6,000 acres of fell. He also donated Banks Intake, Borrowdale (186 acres).[55] In 1958, the year of his death, an agreement was reached with the Forestry Commission not to plant any land at either Butterilket or Black Hall, on land at the top of the Duddon and the two farms were eventually sold to the National Trust. Troutal was bought in 1958 and Black Hall was finally acquired from the Forestry Commission in 1961 (as was Butterilket). Wallabarrow was bought from Lake District Farm Estates in 1974.[56] Baskell, which complemented the Trust's extensive landholding on the Ulpha side of the Duddon, was also given to the Trust by the Symonds Memorial Trustees in 1963.

Another part of Symonds' life work and legacy was the creation of Lake District Farm Estates Ltd which was formed in 1937 as a result of talks between him and R. S. T. (later Lord) Chorley. Chorley and Symonds collaborated powerfully on farm acquisition until Symonds' death in 1958 with Chorley continuing with the task until the late 1970s when the company was wound up. The objectives of Lake District Farm Estates were, "to buy farms in the Lake District and so to keep them that the people of the dales could make a living on them as farmers and that the beauty of their fell and their dale-bottom should be safe."

They were clearly worried not only about the threat posed by the Forestry Commission, but also about the threat from people buying farms in order to be able to occupy the farmhouses for holiday purposes, as was the case with the first farm they bought, High Wallabarrow in 1938.[57] They had other concerns also. They bought Rannerdale in 1938 largely to stop the road up to Buttermere being over-improved and they bought Mireside in 1941 to try to prevent the raising of the level of Ennerdale Water which was being proposed in order to enhance its role as a reservoir.[58]

The original intention of the promoters of Lake District Farm Estates was to try and collect £50,000 over time to invest in farms. The organisation was set up as an Industrial Provident Society which limited the share that any one individual could have in the company. Lake District Farm Estates was particularly interested in acquiring farms in the west of the district where the afforestation agreement did not apply.[59] It also had an important role in acquiring farms before the National Trust was in a significant position to do so. It was quite clear at the time that the National Trust did not have the organisation or the funds to be involved in the systematic purchase of key farms in the 1930s and 1940s. B. L. Thompson, the National Trust land agent, argued that Lake District Farm Estates tapped into resources that the Trust could not tap.[60]

In 1975 Chorley recalled that the founders of Lake District Farm Estates had, "a basic purpose to act as a long stop for the Trust, and in no sense as a rival organisation."[61] The company from the outset saw the importance of not just having the services of a land agent, "to be in charge of buildings and upkeep and to advise on values before purchase," they also gave priority to, "a sheep valuer, well acquainted with local usage and prices, to value on purchase or on change of tenants those local flocks which are commonly the landlord's property."[62]

A few years after Symonds' death, in 1962, when his daughter Susan Johnson produced a revised edition of *Walking in the Lake District,* Lake District Farm Estates owned 3,000 heafed sheep. It was not until 1977 that the company was wound up, when it passed over a group of seven farms or clusters of farms to the Trust. The farms in this group were Gill Bank in Eskdale, High Nook at Loweswater, Mireside in Ennerdale, Howe Green at Hartsop, Harrowhead in Nether Wasdale, Gill, Broadgap and Buckbarrow in Wasdale, Yew Tree and Longthwaite in Borrowdale.

Creating the National Trust Fell Farming Estate
Although the Lake District Farm Estates donation was a large one, the National Trust itself increasingly adopted a strategy of acquiring key fell farms – on some occasions as gifts from the state under the National Land Fund procedures. Harsop Hall was the first farm to be acquired in this way in 1946. Seatoller in Borrowdale was acquired through the same mechanism in 1958 as were Middle Row, Wasdale and Wasdale Head Hall when parts of the Leconfield estates were given to the Treasury in lieu of death duties being transferred to the Trust for "permanent preservation." The landlord's flocks

did not come with the farms and it was sensibly also made a requirement that the Trust should buy the flocks belonging to the farms. National Land Fund procedures also brought to the Trust the nearly 1,500 acre Nether Wasdale Estate in 1964. Field Head in Eskdale, and Underhelm (landlord's flock of 220 sheep) in Grasmere were acquired through the same mechanism in 1974.

Legacies from individuals were also important, none more so than the Goodwin Bequest of 1957 through which the Trust bought Black Hall and Troutal in the Duddon Valley, Butterilket in Eskdale, and Fell Foot in Little Langdale. Row Head was bought at auction in 1962 with a legacy from Mr W. L. Barber and Burnthwaite was bought using the National Trust's own Lake District Funds, generated by appeals, in 1975. The situation in the Duddon Valley was strengthened by the purchase of High Wallabarrow again using Lake District Funds. Brimmer Head at Grasmere came into the possession of the Trust in 1974 and Hoathwaite, Coniston in 1977. Brimmer Head was a particularly important acquisition according to Elizabeth Battrick because of its sheep heafs – which as she puts it, hold the fells at the meeting point between Langdale, Borrowdale, Grasmere and Thirlmere.[63]

There was a steady process of acquisition by the Trust of farms in Borrowdale, a great stronghold of Herdwick sheep keeping. This has led to the situation where the Trust now has all but three of the valley's farms.[64] Seathwaite and Hollows were acquired in 1944, the Nook in 1947, Ashness in 1950, Seatoller in 1959, Stonethwaite in 1964, with (as was noted earlier) Yew Tree, Rosthwaite, with its large landlord's stock of 814 sheep coming as part of the Lake District Farm Estates handover. The three farms and land at Watendlath were acquired at various times between 1960 and 1982.

Various other important Herdwick farms found their way into Trust ownership over the years. For instance, Sir Samuel Scott bought Glencoyne in 1936 when it was put up for sale, in part as building plots, and in 1948 his children passed it to the Trust. The Trust bought How Hall, Ennerdale, in 1950 from the Dickinson family – though interestingly without any landlord's stock, the heafed sheep going with the farm remaining the property of the tenant. In Eskdale, Taw House and Wha House had been bought in 1942. In more recent years, the important Herdwick farms of Bowderdale in Wasdale, and Wilkinsyke and Crag House in Buttermere, have been bought by the National Trust. Townhead at Grasmere was bought in 1981 with its flock of 350 sheep. Fenwick, Thwaites, was bought in the 1990s. Large tracts of fell land also came the Trust's way over the years. Ennerdale Dale has been on a 500 year lease from the Forestry Commission since 1927 (and grazed by sheep from Ennerdale, Buttermere, Borrowdale and Wasdale Head) and Stockdale Moor came through National Land Fund procedures in 1958. The Leconfield Commons, which amounted to 30,500 acres of land and included Kinniside Common, were passed by the State into the ownership of the Trust in 1979.

It is extremely doubtful whether Herdwick sheep farming would have remained such a vibrant part of the Lake District scene if it had not been for the build-up – from all the sources mentioned above – of the National Trust's fell farming estate. The Trust owns 25 per cent of the Lake District. It is the major landowner in several valleys, notably Borrowdale, the Langdales and Wasdale. Of the 91 separate farms owned by the Trust in the Lake District, about 40 have (or should have) landlords' flocks of Herdwick sheep, amounting to a total of 21,000.[65] True, there are plenty of Herdwick flocks which are not kept on National Trust farms and in any case not all National Trust farms require Herdwicks to be kept.

Some of the biggest and oldest flocks are outside the Trust's landholding: for example, West Head which is probably the biggest Herdwick farm of all. West Head is owned by United Utilities – it originally went into the public sector when Manchester Corporation acquired Thirlmere as a source of water in the 1870s. Gatesgarth at Buttermere is a large privately owned farm – though it is subject, as mentioned above, to covenants with the National Trust.[66] Another large, privately owned Herdwick farm is Turner Hall, in the Duddon Valley. It had been owned for many years by the Hartley family and probably has the most consistent track record of success in the show and sale ring over the past 40 years. In addition there are numerous other privately owned fell farms with Herdwick flocks.

But it is the strategic location and critical mass of still persisting Herdwick farms in and around the central dalehead areas (many of which belong to the Trust) that gives this land-holding, and the cultural landscape it represents, such significance. It is not always clear, not least to its tenants and other members of the fell farming community, that the National Trust fully understands the significance of what it has got.[67] For the significance resides, not just as a matter of houses and land, but also as something that has centrally involved the endeavour of skilled local people managing Herdwick sheep flocks over many years; the people who have maintained the landscape of Herdwick country as a vibrant working landscape. There is some feeling in the Herdwick community that this was better understood in earlier days of the National Trust's custodianship of much of Herdwick country. Many would agree with the statement of Bruce Thompson, the first National Trust land agent for the Lake District, when he wrote in 1946 that:

> "Anyone who has thought deeply about the national and local values of the Lake District, and especially anyone who is a native, will feel how important it is not only to preserve this potential national park area as a unique bit of England but to continue so far as possible the local life. Most obviously of all, the sheep farms are characteristic of the region, and so are the men and women who live on them."[68]

Significance of Herdwick Sheep Farming

Those people who began the process of acquisition of fell farms – Mrs Heelis, G. M. Trevelyan, H. H. Symonds and others – seem to me to have understood the significance of Herdwick sheep keeping and the key role of local people in it – and they did much to enable its continuance. They clearly appreciated that this heritage could well have been threatened, on the one hand, by an unbridled free market which might asset-strip even more of the Lake District for its ancient and well-located farmhouses, and on the other hand by the State's need for water and timber. They felt that ownership of the land concerned – under the control of enlightened landlords – was the only true guarantee that the system would continue. They all favoured landscape protection. Symonds and Trevelyan also favoured public access and the recreational use of the fells – and all of them appreciated the need for a prosperous agriculture run by skilled and knowledgeable local people.

It was Symonds who most explicitly linked the landscape and recreation issues to the keeping of Herdwick sheep. In his campaigning book of 1936 against further afforestation in the Lake District a whole chapter was devoted to 'Herdwick Sheep: the injury to local farming.' He argued that they were "a unique breed"; and that Herdwick sheep production was, "the basic industry of the Lake District" on which "the whole economy of local farming depends…" He pointed up the uniqueness of Herdwick sheep breeding. It was "a strictly local industry" with the breed, the technique and the tradition not being found anywhere else. He was particularly focussed on the role of the great dalehead farms – which had the most difficult terrain and which were most susceptible to forestry – and their role in supplying tups for the wider group of farms.

He was concerned to point out the displacement of Herdwick sheep that had already occurred in Ennerdale and round the Whinlatter and Bassenthwaite areas. He calculated the loss to be 3,600 – and he was further concerned by the Forestry Commission's acquisition of two additional dalehead farms at Butterilket in upper Eskdale, and Black Hall at the top of the Duddon. It is clear, however, that he felt that there was a profound ignorance at work – a perspective that did not understand the significance of what the Lake District was, and the role of Herdwick sheep farming in perpetuating it. He argued that, "to some remote official handling files and generalities, from which the breath and detail of life have been squeezed out and dessicated in that facile process from the concrete to the abstract which is a means to all central government, the extinction of a local industry may seem of no account."

For Symonds the particular mattered. The Lake District was a unique landscape and it was, "unique also in those characteristic qualities of life which are bred by man's conflict and co-operation with this landscape." Herdwick sheep breeders were a "unique group of people" in a "unique landscape."[69] He pointed out the "happy coincidence of private profit with public access" – something he did not think occurred anywhere else in the British Isles. This was clearly due to the fact that "our walking on the fells" was "harmless to the sheep," and also due to the realisation of many farmers that the tourists provided an important source of income. The sort of state involvement that Symonds wanted was expressed in his statement that, "The ideal for the Lake District is a national park, not a national forest" with legally confirmed access to and on the fells being a necessary part of that ideal.[70]

Mrs Heelis was sceptical about the potential creation of a National Park in the Lake District. She did, however, favour the acquisition of land by a state-sanctioned and empowered environmental charity, the National Trust. In a letter to the artist Delmar Banner in October 1936, although she had the "deepest respect and admiration" for the institution of the National Trust, she made it clear that the officials of the Trust were "not very satisfying [sic] at buying properties." Banner and his wife Josefina de Vasconcellos were trying to buy a house in the Lake District and were seemingly being priced out of the market. She wrote that they had no chance of buying the Brow in Little Langdale thanks to "proceedings" of money-ed supporters of the National Trust. She continued by saying that, "'Friends' with money have come forward to bid for various lots, including Dale End farm…" She went on to reveal her general position by saying, "I have never been able to fathom what exactly its advocates are aiming at; but I am sure it means interference with other people's property and freedom of buying and selling. And probably an over-visited showplace in another 20 years; it grows more crowded every season."[71]

Mrs Heelis was not at all sure about the policy of encouraging enlightened landowners merely to give restrictive covenants over their land which the Trust began to investigate in 1934 when a large estate came up for sale in Buttermere. Covenants were designed to protect against unseemly development such as caravan parks and forestry and other things destructive of 'amenity'. She felt indeed that the word 'amenity' was over-worked and in a letter to Eleanor Rawnsley in October 1934 she said that she personally mistrusted and definitely disliked "some recent feelers towards a new policy," adding that Canon Rawnsley's "original aim for complete preservation of as much property as possible by acquisition was the right one for the Lake district."[72] One can surmise from this and from the absence of any evidence to the contrary that, unlike Symonds and Trevelyan, she thought little of the emerging English and Welsh concept of National Parks – which has little to say about landownership and land management but much to say about public enjoyment of the countryside.

She was not present at the Annual Meeting of the Herdwick Sheep Breeders' Association in February 1939 at Keswick where there was a discussion on the idea of "converting the Lake District into a National Park." A resolution, proposed by J. F. Buntin and seconded by Jerry Richardson, was eventually carried unanimously which stated:

"That this Annual Meeting of the Herdwick Sheep Breeders' Association views with disfavour any proposal by Friends of the Lake District or any other body for turning the Lake District into a National Park."

The Herdwick sheep breeders seem not to have comprehended the support that the great advocates of national parks gave to their basic economic activity, and seem to have been oblivious of the connection that was being made at a high level of influence between landscape appreciation, outdoor recreation, and traditional farming activities. Mrs Heelis was probably aware of the connections that were being made, but this approach did not enthuse her. Rather as her most recent biographer has put it, "the primary aim of her stewardship was the preservation of a culture of hill farming and sheep stocks."[73] The Lake District National Park was not created until 1951, after Mrs Heelis's lifetime. It was widely felt by advocates of National Parks that many of the problems concerning inappropriate development, including the loss of traditional farming would become a thing of the past. Bruce Thompson, the National Trust land agent, was not alone, when in 1946 he expressed the view that it would largely do away with the need for organisations such as the Trust and others to acquire farm land in order to protect it. He wrote that if a National Park was created in the Lake District, "the National Trust's need to extend its properties would presumably cease within the National Park area because all problems of preservation and access would be settled by the National Park Commissioners."

He hoped that in fifty years time the National Trust estate's acreage would have remained almost static, with what they did have being "models of good taste and good management."[74] With regard to the amount of property, especially farms and farmland, that the Trust now owns, nothing could be further from the truth. The National Trust is now the largest land owner in the Lake District National Park and its fell farming estate is central to the delivery of the area's cultural landscape.

The Lake District National Park

The Lake District National Park came into being in 1951. There was a strong belief in the immediate post-war years that the creation of the Lake District National Park would prevent important property from falling into the wrong hands, and that traditional farming would deliver continuation of beneficial land management. In the early years of the Park Authority (which was styled the Lake District National Park Special Planning Board) this perhaps inadvertently proved to be the case as there was an increasing amount of public subsidy applied to hill farming in the post war years. By the 1970s, however, there were some worrying trends as hill farming areas lost labour and the number of independent farms decreased.

The Park Authority has relatively recently acquired a significant amount of the area's common land by virtue of becoming owner of Torver Commons and the Caldbeck and Uldale Commons (11,300 acres) as a result of the Treasury receiving them in lieu of death duties. But essentially the National Park Authority's approach is one of keeping a watching brief rather than getting involved directly in the ownership and control of land.

Indeed, the Park Authority sold its only hill farm in 2007. So, if the National Park is retreating from its very minor role as an owner of farms, what might it do for traditional fell farming and Herdwick sheep keeping? True, the Park Authority's area rangers can give some support to farmers trying to sort out problems that arise as a result of visitor pressure; the estates teams often provide practical help and their officer who deals with farming and conservation issues can keep a watching brief on developments in the hill farming sector. Unlike the National Park Authority in the Yorkshire Dales, however, the Lake District did not opt to become the vehicle for the delivery of the area's Environmentally Sensitive Area scheme from 1993. This could have made a huge difference and seen a new orientation for the Lake District National Park – one that recognised the importance of its most important land managers. Instead the National Park Authority has been aware of the issues but seems isolated from the farming community.

The Park Authority has not pursued a policy of acquisition of farms. The Authority did, however, by the 1980s own or lease 16,000 acres of common land. It bought Blawith Common in 1970 to add to the various neighbouring commons at Torver (3,700 acres in all) which it leased from the Crown Estate Commissioners. It also managed the 1,055 acres of Glenridding Common and the Caldbeck and Uldale Commons.

The Authority carried out a pioneering project between 1988 and 1990, seeking to try out, on its commons, the prescriptions of the Common Land Forum which had been established by the Countryside Commission. It attempted to see what was required to balance the sometimes conflicting objectives of agricultural use (largely grazing by sheep), nature conservation and recreational access. In particular the project engaged with the farming community and established commoners' associations with a view to produce balanced management schemes.[75]

As it stands, the essential endowment of traditional fell farms is diminishing by the year, amounting now perhaps only to about 100 farms keeping significant numbers of Herdwicks as heaf-going, fell-going sheep. This endowment in the Lake District is made up of the Herdwick farms owned privately, and the National Trust's fell farming estate, pieced together over 80 years and from many sources. The maintenance of Herdwick country depends on this endowment being properly maintained. There are large pressures working against this. As far as the privately owned fell farms are concerned, it is the case that in every valley they have become attractive targets for purchase by people usually not interested in farming them or keeping Herdwick sheep on the fells. There are also doubts as to the National Trust's willingness to maintain fully its fell farm estate. The working landscape of Herdwick country has been for some decades – and remains – under considerable threat as the next section shows. But it also seeks to identify some of the positive developments of recent years and to suggest some prospective lifelines for a cherished cultural landscape.

4: Recent Years: Challenges and Changes

The Rise of the Swaledale

The immediate post-war years were eventful ones for Lake District fell sheep farmers. There was, for some time, a threat that the Labour Government might deprive them of one of their passions by banning fox hunting; there was the Hill Farming Act of 1946 which marked the start of long years of peace-time state support for the sector and there was the atrocious winter of 1947 when many fell flocks were decimated. There was a constant worry in these years about a shortage of hogg wintering – a worry that went on well into the 1950s as Britain struggled to produce enough cereals and other crops to feed itself as the persistence of rationing indicates. There were other concerns too. At the annual general meeting of the Herdwick Sheep Breeders' Association on 23 April 1949 at the Bank Tavern, Keswick, (where it was reported that the Association had 165 members) it was also agreed that a resolution expressing their apprehension that the Lake District might become a National Park be sent to the National Farmers' Union and the Ministry of Agriculture. Later a sub-committee was established to ensure that, "the interests of the sheepbreeders on the fells" were safeguarded.[1]

An outbreak of Foot and Mouth Disease in 1951 led to concerns being expressed about what would happen if it got onto the open fells. It was decided, at a meeting in December 1951, that a letter be written to the Chief Veterinary Surgeon for Cumberland asking, "what was likely to happen if there was an outbreak on a farm which had a fell flock." It was not until February 1954 that a 'long delayed' reply was received. Although we do not know what the reply was, it was accepted by the Executive Council.

A particularly symbolic piece of business reported at the 1949 annual meeting was that the surplus cash balance (£8) from the defunct Haltcliffe Ram Show – which had been a predominantly Herdwick ram fair – had been given to the Herdwick Association – proof positive, perhaps, that virtually all the Skiddaw and Saddleback massif was now given over to the breeding of Swaledale sheep.[2] The keeping of Swaledale sheep was, in fact, becoming very common in virtually all parts of the Lake District.

One of the great ironies of the connection between Herdwick and Swaledale sheep is that whilst the Herdwick Sheep Breeders' Association had, since its inception vigorously defined the differences between the two breeds, the reality was that on the Herdwick farms in the post-war period increasing numbers of Herdwick ewes came to be crossed with Swaledale

Clipping at Gillerthwaite, 1930s.

rams. Swaledales, especially since the end of the Second World War, have become the dominant hill sheep of the north of England. Official estimates suggest that in the early 1950s there were about 200,000 Swaledale breeding sheep in Cumberland, Westmorland and Lancashire (with a similar number being found in other parts of England) as against the 75,000 Herdwicks nationally which (with the exception of 101 in Yorkshire) were entirely found within Cumberland, Westmorland and Lancashire.[3]

The majority of Herdwick farms in the post-war world began to use a few Swaledale tups alongside their Herdwicks. Not only was there some economic gain from cross-breeding but it was also the case that Swaledale tups were now easier to get hold of as transport improved and it no longer seemed a particularly long way to Kirkby Stephen and Hawes, or even to Middleton-in-Teesdale or St John's Chapel. Initially it was favoured because it was a bigger sheep than the average Herdwick and like all first crosses the offspring produced had hybrid vigour. It became very common from the 1950s onwards on most Herdwick farms to have a growing number of sheep with spotted bodies and small spiky horns within the fell flock. They may have looked like mongrels but they were capable of performing well. Over time, and with repeated crossing with the Swaledale tup many a Lake District Swaledale flock has been established. True, some die-hard Herdwick breeders would have nothing to do with them and even preferred to experiment with Welsh Mountain and Cheviot sheep rather than entertain the Swaledale – which rapidly became the dominant fell sheep of the Lake District.

The biggest acceleration of the infusion of Swaledale blood into the Herdwick flocks came from the 1960s onwards with the development and marketing of an outstanding livestock product, the North of England Mule (the product of a Blue-faced Leicester tup and the Swaledale ewe). The Mule soon assumed its subsequent position as the supremely economically successful lowland ewe. As is sometimes said of this development, "the Swaledale made the Mule and the Mule made the Swaledale."

Since people needed extra income, increasing numbers of Swaledale tups were used in Herdwick flocks to produce draft ewes for sale or horned semi-blackfaced ewes that might breed some thing resembling a Mule lamb. In the late 1960s and early 1970s there began to be, for instance, a marked difference between the returns from the typical Herdwick farms of the Lake District as against the Swaledale farms of north Westmorland, as represented by the prices of stock at Troutbeck and Kirkby Stephen auction marts. The differential was between 30% and 100%. In 1970, for instance, Herdwick ewes averaged £2.35 at Troutbeck, whereas Swaledale ewes averaged £6.60 at Kirkby. Even Swaledale ewes at Troutbeck were little more than half the value of Swaledales at Kirkby – no doubt because they were probably recent crosses from the Herdwick.[4]

Throughout the period, however, there were many Herdwick loyalists who stuck to the breed, but regrettably many good Herdwick stocks were lost in the 1950s and 1960s, and increasingly Herdwick country retreated to the western and central fells. In 1957, for instance, only 190 Herdwick shearling rams were registered by the HSBA compared to the 350 per year registered

Herdwick ewes at Millbeck, Great Langdale, from the flock of Mr J. V. Gregg, 1964.

over the past few years. Although the HSBA concerned itself to some extent about new developments and promotional activity, for instance by being involved in trying to develop a Herdwick tweed in the early 1960s and by attending various promotional events throughout the North of England, generally speaking it was on the decline. There were, however, a number of enthusiasts who concentrated on the show ring.

It was, in general, a period of serious competition and great rivalries with big turnouts at shows throughout the season. There were also celebrated sheep such as Ernest Tyson's tup Dick's Permission (so called because it was the result of a mating between a ewe of Ernest's and Dick Wilson's tup Perivale – a liaison for which Ernest had to get permission). Another renowned tup was Derwent Tyson's Blencathra Fenwick.[5] By the 1960s the Wilsons of Herdwick Croft showed a lot less but breeders such as Derwent Tyson, Thomas Richardson, John Richardson, Ernest Tyson, William Bowes and Tyson Hartley remained active into the 1970s and 1980s and, in some cases, beyond.

New enthusiasts emerged from the 1970s such as Harry Hinde, Gilbert Tyson, Gowan Grave, Teddy Tyson, Syd Hardisty, Harry Hardisty, Joe Weir, Kathleen Weir, Vic Gregg, Arthur Weir, Harry Robinson, Dick Wilson, Jean Wilson and Harry Wilson. These people and others, fought it out with their sheep at shows throughout the Lake District and sometimes beyond. Gowan

Eskdale Centenary Show, September, 1964. Male and female Champions, bred and exhibited by W. D. Tyson and Sarah Tyson. Judges (left to right) stand behind: J. Richardson, S. Edmondson, P. Temple, R. Teasdale and M. Askew.

Dick's Permission, owned and bred by E. Tyson and Son, Broadrayne, Grasmere. Winner of 97 First and Champions and 13 Reserve Champion prizes.

Grave in particular supported the big shows like the Royal and the Great Yorkshire, as well as being an ever-present exhibitor over a forty year period at local shows. At the Royal in 1977 a tup of his won the prize for the champion sheep of all breeds. An equally committed stalwart over six decades was Tyson Hartley whose Turner Hall flock bloodlines have been very influential in the breed.

Annual Meetings from the 1950s onwards had usually less than twenty people present and there were some years when the secretary found it difficult to get round to inspect and register rams. At the 1972 AGM Gilbert Hartley announced his intention to resign, "partially because of problems of registration." Hartley had replaced William Tyson as secretary in 1954 and he served in that position for twenty years. There were the inevitable loss of stalwarts every year but in some years they were replaced by significant numbers of new members. At the AGM in 1972, for instance, it was reported that there were five deceased members but there was also a large group of over twenty new members. Hartley did not relinquish the secretaryship until 1974 when Joe Cowman took his place. In the same era George Wilson of Glencoyne energetically took on the role of chairman of the Council and in the 1970s there was active participation in events organised by the National Sheep Association and other bodies. The Council was replaced by an executive committee in 1982 – with Teddy Tyson of Broadrayne, Grasmere, taking over the chairmanship shortly after. Joe Cowman died in 1984 and was

Geoff Edmondson loading sheep at Naddle (top of Haweswater).

replaced as secretary by William Bland after the latter retired from Woodend, Ulpha. He did the job until 1990 when the present writer took over.

On the farms themselves an economic pragmatism set in, and in these years a large number of fell farms developed a Swaledale flock (bred out of Herdwicks) alongside the Herdwick flock. By the 1980s there were very few large-scale Herdwick flocks outside of the central, western and south western dales of the Lake District. By the end of the 1980s, for instance, what had been the last remaining, and one of the largest Herdwick sheep farms on the eastern side of the Lake District, that at Naddle, which ran several large stocks of sheep in Mardale and round Haweswater, had put everything to the Swaledale tup for a number of years and only a few long-lived Herdwicks remained on the hardest heaf.[6]

A breeder interviewed at the Keswick Tup sale in October 1989 admitted that he, "slipped a few more Herdwick ewes to Swaledale tups every year," and others bemoaned the differential between the value of Swaledale and Herdwick sheep. The wool price was low, with the great bulk of the previous two years' clip being unsold, although a little had made a mere 29p a kilo (which would now be regarded as a good price). Wool Board member Bill Rawling from Ennerdale warned that there "may be doubts about the future of the Herdwick as a breed kept in commercial numbers" and he drew attention to the fact that this would represent "the loss of an important part of the Lake District's heritage."

The National Trust at that time was conducting a campaign about the breed and in a North West Newsletter they had a headline about Herdwick sheep which read, "These sheep need help" with the Trust's regional information director suggesting that if recent trends continued then "the breed will disappear." The breed association secretary William Bland nevertheless put a brave face on it and pointed to reasonable prices at both the Keswick and Broughton tup sales and also the fact that small flocks were being created all over the country[7] – a point that had been discussed at the HSBA executive committee in 1984-5 – though this was later to prove to be more of a problem than a blessing as what might be called the 'Beatrix Potter' effect grew in significance over the years.

From the 1980s the Herdwick community increasingly found itself involved in a wider world of harsh realities. Remote Lake District fell farms found that they were increasingly linked to global events. Lambing time of 1986, for instance, coincided with the accident at the Chernobyl nuclear plant in what was then the Soviet Union. The fallout from this accident hit the ground in, among other places in Europe, the western Lake District, contaminating the grazings of 1,670 farms, in an area close to the Sellafield nuclear plant. This seriously disrupted movements and marketing on a group of farms from Loweswater to the Duddon Valley. Restrictions were still in place on nine Cumbrian farms involving 11,500 sheep in 2003 and they could remain until 2035.[8] The events in Tiananmen Square in Beijing led to China being unable to get the credits it needed to remain a big buyer in the international market for low value wools, which led to even lower prices for Herdwick wool. The onset of BSE in cattle was later to lead to concerns

Two-shear tup, breed record price, 5,600 guineas, Cockermouth, 2004. Bred by Stanley Jackson, Nook, Rosthwaite.

about scrapie[9] in sheep and as a permanent backdrop to everything there was a continuous process of reform of the Common Agricultural Policy.

In spite of all the turmoil of these years of uncertainty it was evident from the 1980s, however, that there was still real enthusiasm for the breed. From the 1980s younger breeders were the driving force behind the establishment of new opportunities for showing sheep which had a more workaday life than those reserved for the show circuit. Shepherds' Meets shows with rules debarring any sheep which had been shown at agricultural shows in that same season enjoyed a resurgence. The long established Shepherds' Meets and shows at Wasdale Head and Walna Scar also attracted more interest. Buttermere Shepherds' Meet show was established in the 1970s. In 1988 in Borrowdale, the old Rosthwaite sheep fair which had not been held for many years, was revived in the form of Borrowdale Shepherds' Meet and show.

In recent years there is continuing evidence of the pre-eminent suitability of the Herdwick for the climate and terrain of their heartland – the central and western part of the Lake District. Farms often experimented with a few Swaledales for a number of years, before realising that they did less well on sparse grazing, needed supplementary feeding and struggled in difficult winters and slow springs. They also needed to be drafted from the fell earlier. Simon Keld, on Kinniside Common, for instance, was one of those farms, a point made by Sam and Jack Wightman who farmed it until the early 1980s. It was also the case at Hows Farm in Eskdale. Noel Baines told me a few years ago that if they had carried on keeping Swaledales any longer they would have had no sheep left.

There seems to be little doubt that the Herdwick is a locally adapted sheep par excellence and that it is most fit for its purpose on the high, hard and wet fells of the west and central part of the Lake District. There has been a consolidation of the number of farms breeding Herdwicks over the past twenty years – though of late there has been a marked reduction in the number of Herdwick sheep due to agri-environment schemes, the effects of the handling of Foot and Mouth Disease, and more recently the general effect of the decoupling of support from production.

Years of Expansion: The Numbers Game

For a period of over fifty years since the end of the Second World War hill farmers have been given incentives to bring about the general expansion of production. Subsidies were available for hill farmers to expand their flocks and herds, and they were encouraged by policy transmitted on the ground by agricultural extension workers to adopt new technologies and to apply new knowledge. The Agricultural Development and Advisory Service (ADAS) had a powerful role in informing farmers about technical efficiency and made grants available for modernisation.

The Agriculture and Horticulture Development Scheme (AHDS)

incentivised farmers to extend their enterprises, often considerably beyond the capacity of the home farm to sustain them. For instance, a small hard fell farm like Mireside, at Ennerdale, in 1980 under the AHDS ran 823 ewes and shearlings plus 220 gimmer hoggs alongside twenty suckler cows, one bull, eighteen bucket reared calves and fifteen tups. The landlord's stock on Herdus was only 136 ewes, 30 gimmer shearlings, 35 gimmer hoggs and two tups – about double or treble which (with a small number of cattle) would probably have represented the true carrying capacity of Mireside's inbye resources.[10]

The essential ingredient was that, as the size of herds and flocks increased, the amount of payment forthcoming from the Hill Livestock Compensatory Allowance also increased. Farmers themselves spoke disapprovingly of this being 'the numbers game' – something they were concerned about, but also something which it would have been folly to have resisted. There were no upper limits on support in England and Wales (unlike in France and West Germany), and there was no distinction in the Severely Disadvantaged Areas about the degree of handicap (between say the Lake District and Snowdonia on the one hand, and Exmoor and the easier parts of the Peak District on the other).

Generally all this had the effect of rewarding the bigger and better off farmers more than the smaller or more difficult farms. These were also years of considerable price inflation – enabling expansion-minded farmers to buy additional land for sums which relatively quickly became reasonably easy to service. The trend towards expansion and amalgamation was thus encouraged both by the state and by the market, but it led to ever smaller numbers of people managing larger amounts of land.[11]

It was estimated in the early 1970s that there were only seven people involved in farming in Grasmere, eight in the Troutbeck valley and fourteen for the full length of Mallerstang, with the inevitable consequence of deterioration of landscape maintenance. A detailed study of Hartsop by the Countryside Commission in 1976 concluded that neither the Lakes Special Planning Board nor the local authorities had the power to tackle the problems and challenges of rural housing, the plight of the small farmer, and the maintenance of the landscape. The Planning Board's National Park Plan of 1978 concluded that if existing trends were allowed to continue there would be "a sustained loss of farmers and farm workers" which would result in "the continued deterioration of the landscape and add to the problems of rural depopulation."[12]

The Hill Livestock Compensatory Allowance did, however, embed good management practices by specifying that 'hardy breeds' (of which Herdwicks and Swaledales were examples) had to be used to qualify and it was a requirement that flocks were managed in 'regular flock ages,' i.e. there was a presumption that hill farmers would breed their own flock replacements, a practice which has clear animal health benefits. The ending of the UK's sheep 'variable premium' which 'made up' the market price for sheep meat to an agreed amount, led to very low market prices until it was replaced with the EU approved Sheep Annual Premium.

This system improved matters in the medium term but in the transition period things were very difficult. Some calculations of gross margins on a typical central Lake District Herdwick farm carrying 700 ewes and 25 suckler cows, comparing 1988-89 figures with those for 1990-91, showed a reduction in net farm income from about £9,000 to just over £3,000.[13] Market prices continued to plummet in the rest of the 1990s. The Ministry of Agriculture Fisheries and Food calculated that the average net farm income for sheep per Severely Disadvantaged Area farm in the UK in 1995-96 was £14,900, reducing to £8,600 in 1997-98 and to £3,500 in 1999-2000.

There were, however, environmental consequences of expansionary policies as the conservationists were to point out repeatedly. The productionist policy of "food from our own resources" (to use the title of a White Paper of the late 1970s) led, in some places, to over-grazing and to serious concerns about the condition of many fells. These concerns were raised most significantly for the first time in Pearsall and Pennington's book, *The Lake District* in the Collins New Naturalist series, first published in 1973. Pennington wrote of the grasslands of the fells that, "Over-grazed pasture is now extremely widespread on the Lake District fells, both on the highest intakes and on the unenclosed fell commons above them."

D. A. Ratcliffe contributed a special study of the Skiddaw Forest area to that volume and attributed a loss of heather to "an increase in the stocking density of sheep."[14] Ratcliffe, who was Chief Scientist to the former Nature Conservancy until the late 1980s, and who in 2002 produced his follow-up to the Pearsall and Pennington volume, had much more to say in retrospect on over-grazing. He felt that headage payments for hill sheep had promoted "relentless over-stocking of the fells." He felt that what sheep reductions there were through ESA and other schemes were achieving little, with attendant deterioration of heather and bilberry heath, because the reductions were insufficient.[15]

The Sheep Annual Premium, which existed from the early 1990s until the arrival of the Single Payment Scheme in 2005, had the effect of keeping production at high levels even when markets were poor. This was done through the use of a system of sheep quotas whereby farmers were paid on the number of sheep they retained over a substantial part of a year. In general, however, the last ten years have been challenging ones. Hill farm incomes have declined dramatically and there has been a difficult external environment with a mixture of positives and negatives, namely the arrival of the Lake District Environmentally Sensitive Area (ESA) and other agri-environment schemes; the Foot and Mouth Disease (FMD) epidemic of 2001, and (as a general background of change) continuing reform of the Common Agricultural Policy (CAP). Each of these is now dealt with in turn and the connections between the three themes are explored.

The Agri-Environment Challenge

The biggest general challenge has been that of nature conservation. A growing body of opinion developed arguing that the provision of incentives to increase production in the hills had done serious damage to the flora of the fells. The Lake District has had an Environmentally Sensitive Area scheme since 1993. It was designated in the second wave of ESAs at a time when market incomes were very low. The Pennine Dales ESA, which included parts of Cumbria, was designated in the first wave. Other farms in upland Cumbria were eligible only for the Countryside Stewardship Scheme if they had the right target habitat.

The Lake District ESA was run on a whole farm approach and included fell land, although initially it largely concerned just the in-bye. This was due to the fact that it did not prove possible to secure full agreement from all commoners on the large area of common land in the Lake District. But it was eventually recognised that not all rights holders had to agree in order to have a viable scheme including fell land. In some places also it was possible to split commons into different grazing units. With Buttermere common leading the way, other commons followed suit with eventually 80% of common land in the Lake District becoming part of the scheme.

The Lake District ESA was a large programme which eventually had 170,000 hectares under agreement. The scheme reached a peak of expenditure around its tenth anniversary when £10 million was disbursed. Because the scheme was area-based, some farms earned very large payments which, when the money was spent on the buying or renting of land, had a distorting effect on the market. But the Department for Environment, Food and Rural Affairs (Defra) was interested in other outcomes. A Defra briefing note produced after the first ten years of the scheme indicated that about 1,000 farm buildings were renovated, over 120 miles of walls rebuilt, and 150 miles of hedges renovated.

Essentially the scheme rewarded good practice and was regarded as having "raised the environmental profile for the farming industry" in the Lake District and as having "changed attitudes towards environmental issues and a more sustainable approach to agriculture." Over 800 hectares of flower-rich meadows and 4,500 hectares of herb-rich pastures and valuable wetlands were brought under agreement. On the fells it was estimated that an overall reduction of about 25% of the sheep flock had been achieved, contributing to "more sustainable levels of grazing."

But there were also concerns at this time that the ESA was often paying for maintenance rather than enhancement and more attention began to be paid to the targets for improving the condition of SSSIs (Sites of Special Scientific Interest). In some places English Nature provided additional funding through the Wildlife Enhancement Scheme and only recommended renewal of agreements which involved SSSIs and led to further reductions in sheep numbers and active management to improve habitat quality. This seems to have weakened the traditional heafing system especially where flocks were completely off-wintered. More recently, however, a significant number of grazing commons have been entered into ESA schemes. This phenomenon has become more widespread especially after Foot and Mouth Disease in 2001 when reduction of numbers through culls has meant that income needs to be replaced and achieving lower stocking rates is no problem. Since FMD, for instance, Kinniside and Nether Wasdale Commons have entered into ESA agreements involving big cuts in stocking densities in both cases – and some of the Ulpha commoners have also considered it.

The consequence, of course, is that when sheep flocks get smaller (due to whatever cause) the heafing system is being disrupted as sheep move into territory formerly occupied by other flocks. If cuts in numbers are too large, sheep spread out too far – almost inevitably into the next valley – management becomes extremely difficult and at worst the heafing system breaks down. There are now considerable additional costs in time and money of retrieving strays from distant farms. For example, Borrowdale farmers go to fetch stray sheep round by road from upper Eskdale, a round trip of just over 100 miles, every time the sheep are gathered in for one of the regular processes of sheep management (tupping, lambing, clipping, dipping, weaning of lambs and drafting of older ewes). Although some ESA agreements on fells were renewed before the scheme was replaced and thereby extended, in all cases over a period of time they will be replaced with Environmental Stewardship which is only available at higher level to commons of above average environmental interest. Fortunately however there will be wide access to the Upland Entry Level Scheme (UELS) for commons and a supplement of £5 per hectare per year.

Another unintended consequence of these cuts in numbers has been the increasing incidence of ewes bearing twin lambs. Describing a general problem, Anthony Hartley of Turner Hall, in early 2008 stated that the result of the 45% cut in the stocking rate on his holding was a substantial increase in fecundity. About 30% of his ewes are now having twin lambs. The number of twins is rising year on year and this is creating management problems. Twin lambs cannot, on nutritional grounds, be put to the fell with their mothers until they are well established physically (for instance after the ewes have been clipped) and so this means that a significant proportion of the replacement flock is not well heafed because of spending only a few weeks of their first year at the fell. Ewes with twins need managing during the late spring and summer on the lower ground of the farm.[16] Needless to say not all farms have this spare grazing capacity and financially (and environmentally) costly summer grazing in lowland areas has to be taken.

Foot and Mouth Disease 2001

The world's largest ever outbreak of foot-and-mouth disease occurred in the UK in 2001. It struck hardest in Cumbria – which had 44% of all 'Infected Premises' and where nearly 2,000 separate farm holdings lost some or all of their stock in culls. The stark statistics are that 1.25 million sheep and a quarter of a million cattle were culled in Cumbria.[17] Although the majority of the actual cases were in lowland Cumbria, especially in the north of the county, and therefore outside Herdwick country, many Herdwick sheep were lost in the lowlands where they were wintering. About a dozen Herdwick sheep farms were completely or heavily culled due to FMD. The majority of breeders, however, lost one or more generations of young female sheep away on their wintering grounds. It is estimated that 30% of the breed's usual 60,000 strong ewe flock were lost in FMD culls. In addition, a significant number of tups (many of them highly resistant to scrapie as a result of a major genotyping programme over the previous five years) were lost.[18]

The discovery of FMD at Black Hall, Ulpha, (a renowned Herdwick sheep farm at the top of the Duddon Valley) in late March 2001 led to huge worries amongst the fell sheep farming community of the Lake District about the consequences for heafed (or hefted) sheep in general and Herdwick sheep in particular. The central fells of the Lake District are arguably one big grazing area and, it was feared, given the inter-mingling of flocks on the fell-tops there was considerable potential for the spread of infection to all points of the compass. It was feared that both the threat and the reality of even scattered infection might lead to a draconian process of culling.[19]

Fortunately, the Black Hall case, and other cases lower down the Duddon Valley on the Ulpha side, did not lead to the spread of infection into the southern part of the central fells. However, the heavy culling of Swaledale flocks on the Skiddaw massif in the northern Lake District led to great concerns that a similar approach might have been adopted in the central and western Lake District – i.e. in Herdwick country itself. From late April and early May 2001 a different FMD management policy and practice (based more on observation and blood testing rather than draconian culling) began to apply, recognising the serious issues concerning special breeds such as the Herdwick and the Rough Fell and their greater vulnerability in, and their particular contribution to, the Cumbrian fells.

Many organisations, in addition to the breeders themselves, began to point out the value of the role of the heafed flocks of Herdwick sheep in the management and maintenance of the Lake District landscape. These included the Countryside Agency (which along with the Cumbria Community Foundation and Barclays Bank gave some financial support to the Herdwick Sheep Breeders' Association to help get it through the crisis), the Lake District National Park Authority, English Nature and the National Trust. Many breeders were surprised at the support which came from bodies which had previously seemed antagonistic by virtue of their always seeming to focus on over-grazing (and therefore the removal of sheep from the fells) rather than the positive role of grazing. Concerns about over-grazing tended to evaporate (at least in public) as the more basic issue of how grazing would be maintained came to the fore.

There seemed also to be a high level of consensus about the importance of Herdwick sheep keeping and the heafing system as being particularly congruent with high levels of public access and that this was brought about long before the bringing in of the Countryside and Rights of Way (CROW) Act. In the Lake District there were already some circumstances which legally allowed public access to high common land, i.e. under the local government legislation concerning commons in Urban District Council areas such as Lakes UDC which included the Langdale and Grasmere fells amongst others. There was also legal access to common land owned by the National Trust.[20]

It was, however, the heafing system in the Lake District which had for many years enabled England's best mountain scenery to be largely unfenced: heafing facilitated open access to the country's highest fells. This endowment of the greatest outdoor recreational resource in England is very largely a product of common land and open-fell sheep farming and the heafing system. The scale of this, and the absence of grouse shooting, makes the Lake District different from the Yorkshire Dales or the North York Moors. A great part of what is arguably the most important of the Lake District's 'public goods' have been produced by generation after generation of fell farming activity. Certainly, by the time the Cumbria FMD Inquiry reported a year from the approximate end of the outbreak, it was widely accepted that there was an economic vulnerability of the cultural land heritage of upland Cumbria, related to the common land system and the used of heafed flocks. Support for additional shepherding and the appropriate use of fencing were essential if heafed flocks were to be regenerated.[21]

One of the consequences of FMD culls was that fell farmers who lost substantial numbers of sheep were either unable or unwilling to restart or rebuild their fell flocks unless there were fences round their heafs. Fencing was carried out at Buttermere to stop encroachment of Ennerdale sheep into an area depleted of sheep (which would have prejudiced an agri-environment agreement agreed at the end of FMD) and also at Ulpha to protect the heafs of the farms at Woodend, Baskell and Hazel Head – all of which had lost their sheep which were at the fell in early 2001. In a few cases neighbours on commons sold surplus heafed sheep to farmers who had lost theirs, giving them the opportunity to rebuild. Most commonly, however, where culls had been of one or two generations rather than effectively the whole flock, the predominant technique of the Herdwick breeders was to continue to put the older ewes (which normally would have been put to a terminal sire tup to produce a cross-bred lamb) back to the Herdwick tup for another one or two seasons. Income foregone as a result of this necessary management option was considerable, but breeders were keen to get their flocks back up to strength and, where possible, to optimise their numbers in order to access additional agri-environment income.

English Nature and FMD

English Nature took the opportunity of there being reduced sheep numbers on the fells, to see if these could be fixed through agri-environment agreements involving reduced stocking. The organisation regarded this as a well-meaning response to FMD, though it was one which the organisation at the time communicated badly, giving the impression that it was driven by the opportunity that was presented by the reduction of sheep numbers in many places, rather than by a sympathetic response to the issue of sustaining livelihoods in the hill farming community. Between 2002 and 2005 English Nature and Defra secured agreements with graziers over about 46,000 hectares and involved payments of £4.7m to farmers who complied with grazing their fell land at what were felt to be sustainable levels. The schemes involved flock reductions, over-wintering, shepherding and tailored agreements. In some areas it was possible to intervene at a landscape scale using a 'whole fell approach,' for instance on the Helvellyn and Skiddaw massifs.

In 2005 English Nature wrote up what it called, "The sustainable grazing initiative in Cumbria, 2002-2005." The common goal of the projects which made up this initiative was about ensuring that, "the designated upland wildlife sites in the county are grazed in a way that maintains or restores their value for wildlife." To achieve this involved, "shifting to suitable stocking levels that maintain or restore upland vegetation to good condition." The rationale revolved around the broad changes in agriculture; the need for English Nature to make a positive response to FMD and the key priority of improving management of SSSIs. They showed that their response was rapid in areas worst affected by FMD and stated that, "It would not have helped farmers if English Nature had let them proceed with large-scale restocking, and then late in the day come to discuss proposals to reduce stocking levels."

It must be commented here that this well-meaning interpretation did not always find favour in the hill farming community. Rather it seemed to many that this was part of a deep laid plot about seriously reducing the number of sheep in the county. The English Nature intervention was regarded as a way of tidying up some of the loose ends after the heafing system had been fundamentally weakened. English Nature defended itself against accusations of their interventions causing a weakening of the heafing system, by saying, firstly, that the heafing system had been in decline for some years on some fells and, secondly, that some heafs only worked because of high stock numbers – the pressure of adjacent animals keeping flocks in position, "rather than a learned sense of place on the fells," indicating that they actually took rather a mythical view of heafing.

Heafing does depend on learnt knowledge of the grazing areas but this is mitigated by fell sheep naturally moving into fresh grazing areas if they are available. They did accept that stock reductions might well make this worse, "reducing grazing can create a vacuum into which stock move." Given that the prescription was normally to reduce the stocking rate a further 40% from the ESA level, they were inevitably forced to concede that stock reduction of this order did make gathering more difficult and did have knock-on effects. They even acknowledged that they were responsible for a good deal of the increase in Herdwick ewes which were available at the draft ewe sales in 2004 over 2003.

Although English Nature seemed to understand the problems, they essentially took a defensive viewpoint being unduly concerned about how much of the damage to fell farming systems had been caused by their interventions. It might have been more widely appreciated in the fell farming community if they had expressed concern for the integrity of the system as a whole and all the pressures it comes under. These pressures range from CAP reform, to issues of low profitability and to the sheer lack of people available to carry out the necessary management tasks. It is unhelpful to make interventions which just have one focus such as plant bio-diversity when in reality what is involved is multi-faceted. They conclude the report by saying that the agreements have made, "a significant difference to the condition of upland SSSIs in the county and encouraged a way of farming with considerable environmental outcomes including keeping internationally important wildlife sites in favourable condition." In reality there is much more at stake than vegetation.

It is to be hoped that the new government body in charge of all this since 2006, Natural England, will fully recognise in its local practice the importance of the contribution of hill farmers. The new agency has as its purpose, "to ensure that the natural environment is conserved, enhanced and managed for the benefit of present and future generations, thereby contributing to sustainable development." One hopes also that its statement that this cannot be achieved without 'partnerships' including with 'land managers' will really ring true with its implications fully comprehended. There should be much more to this agenda than the delivery of incentives to de-stock the fells.[22] The people with the skills and commitment to manage grazing animals on the fells need to be centre-stage.

The National Trust and the Foot and Mouth Crisis

During the FMD epidemic of 2001, other than the efforts of the Herdwick breeders themselves, the National Trust probably did most to raise awareness of the threat that was posed to the Herdwick sheep breed, their landscape, and the community which rears them.[23] Oliver Maurice who was regional director of the Trust at the time took a leading role in dealing with the immediate crisis issues[24] and in campaigning for a recovery programme. While the FMD storm was still raging its way through Cumbria, the Trust produced a document, *A Vision for the Lake District after Foot and Mouth.*

The document seems to show that, as a result of FMD, the Trust gained a clearer strategic understanding of the importance of their land holding and its cultural, economic, environmental and also social values. The vision document argued that the FMD outbreak was proving to be a 'watershed' for the Lake District. It had brought into focus the "decline in the viability of the rural economy" and had brought into the open a debate about how to sustain "one of the most critical and valued living landscapes in Europe." Its purpose was to steer collective efforts, "to drive forward the restoration of a sustainable, local economy post foot and mouth." The three most important things were, "its spectacular cultural landscape, its massive tourist potential and its vulnerable upland farming" – all of which were linked. It listed social, environmental, economic, and cultural issues and suggested ways in which, through partnership working, they could be addressed. Local housing issues, the development of infrastructure for the local produce sector, the adoption of more sustainable farming practices and the investigation of World Heritage Site cultural landscape designation might help: these were the 'big ideas.' A number of pilot schemes were suggested to take them forward: including farming careership schemes, the carrying out of whole farm plans, local processing initiatives, developing outlets for Herdwick products and appointing a Herdwick project officer.[25]

The FMD crisis of 2001 revealed many of the weaknesses of the Lake District economy, especially parts of it which were central to the National Trust's mission. Thousands of breeding sheep were lost, a relatively large number of fell farm tenants gave notice of their intention to retire, and some others moved to better farms in other areas. The whole Herdwick inheritance seemed to be threatened. At one point the *Westmorland Gazette* had a front page headline proclaiming, "Trust tenants desert the fells." Oliver Maurice, realising the challenge involved in re-building the fell-going stocks, and the impact this would have on already shaky economic prospects, stated at several crisis meetings that the Trust might have to have 'negative rents': i.e. that they might have to pay people to run fell farms at least in part if they wanted full maintenance of their farms' public goods – the landscape and countryside features, bio-diversity as well as the cultural landscape of properly managed heafed, fell flocks.

Letting particulars for Trust farms now clearly state the Trust's purposes as an "environmental organisation." Using as an example the particulars for High Wallowbarrow in the Duddon Valley where a new tenant took over in late 2003, it was stated in the letting particulars that applicants ideally needed to have experience of managing a farm of this type ("a hard and remote hill sheep farm.") Good experience is needed in animal husbandry, "particularly with upland sheep and the Herdwick breed, together with experience and an understanding of the management practices of shared Common grazings and gathers and working with neighbours is also important." It is also,

> **THE NATIONAL TRUST**
>
> **NOTICE**
>
> **ULPHA COMMON**
>
> District of Copeland - County of Cumbria
>
> NOTICE is hereby given of the intention of The National Trust (i) to ERECT TEMPORARY ELECTRIC FENCING on a portion of the above-named Common of an initial length of 1870 meters, surrounding and subdividing 75 hectares into 45 hectares and 30 hectares, with access gates, from March 2002 and extending in September 2002 to 6490 meters surrounding and subdividing 168 hectares into 90 hectares and 78 hectares to remain in place until March 2006, on Hesk Fell and Cockley Moss for the purpose of containing non-hefted sheep on the common following Foot and Mouth Disease. (ii) to ERECT TEMPORARY ELECTRIC FENCING on the Common of a total length of 2450 meters surrounding 46 hectares, together with access gates and stiles, from March 2002 until March 2006, situated east of Rough Crag and south of Great Worm Crag, for the purpose of containing non-hefted sheep on the common following Foot and Mouth Disease; and (iii) to apply under section 194 of the Law of Property Act 1925, to the Secretary of State for Environment, Food and Rural Affairs for consent to the erection of the proposed fences and access gates.
>
> A map showing the position of the proposed fences may be inspected at Ulpha Post Office, Ulpha, Broughton-in-Furness, between the hours of 8.30am and 5pm (excluding the hours of 12.30 pm to 1.30 pm) from Monday, Wednesday, Thursday and Friday each week, and between the hours of 8.30am and 12.30pm on Tuesdays until the 10th day of December 2001.
>
> Any objections or representations relating to the proposed temporary electric fencing should be submitted in writing to THE SECRETARY, DEPARTMENT FOR ENVIRONMENT, FOOD AND RURAL AFFAIRS, ZONE 1/05, TEMPLE QUAY HOUSE, 2 THE SQUARE, TEMPLE QUAY, BRISTOL BS1 6EB ON or BEFORE that date quoting reference CYD3.
>
> For and on behalf of The National Trust, North West Region
>
> R PALMER
> Property Manager
> The Hollens
> Grasmere
> LA22 9QZ

Ulpha Common notice from the Foot and Mouth outbreak, 2001.

importantly in my view, made clear that it is an "economically part-time farm." Since FMD, however, there has been some strong criticism from the fell farming community of the National Trust's custodianship of its Herdwick farm estate.

The National Trust is very much of the view that many of the problems they have can be put down to reform of the Common Agricultural Policy. In a document called *Impact of Common Agricultural Policy Reform on the English Uplands* (2006), it indicated that:

"There is now a real threat of farmers not keeping cattle and sheep because they cannot make a viable living from farming. This may lead to chaotic and unplanned loss of land management capability with potentially severe and widespread consequences."

The Trust study investigated the situation on 60 tenanted farms in the Lake District, the Yorkshire Dales and Moors, the Peak District and Northumberland. The analysis showed that by 2012 most of the farms will be making a loss. In some cases the loss might be more than £10,000. Whilst the Trust study welcomed the CAP reforms because of the significant environmental benefits through appropriate grazing levels that they will bring, they feel that without further changes some farmers may get driven the other way with the result that important areas of our uplands are no longer grazed. They call for, "a managed process of transition rather than the faster and unplanned change in prospect."

The study suggested that, "It would be far better to manage future change by maintaining appropriate livestock systems in the short term, which will maintain the environment we want rather than incur the costs of having to reintroduce them at a later date." It called for Government to have a vision and a clear strategy for the future of the uplands. It wished in particular to see the development of the Hill Farm Allowance so that it could be used in a more flexible way to underpin businesses already delivering environmental schemes and public goods, and suggested that the budget for it should increase from the current £27m to £50m. It also called for an advice service for farmers who wished to develop their businesses and skills, and to help those who wished to retire or leave the industry. It suggested that there be more investment in the transition from production-based agriculture to environmentally sensitive land management over the next five to ten years.

The Trust's message is, in most respects, exemplary and well evidenced and its voice is a powerful one, for instance, on Defra's Upland Land Management Advisory Panel (ULMAP). But some of the Trust's practice on the ground in the Lake District has left much to be desired. Its future policies will certainly not be the right ones if it continues to use arguments about the inherent future unviability of many of the fell farms as a signal to carry out more closure of farms with subsequent re-parcelling of land to neighbouring farms as was recently done at High Yewdale, Coniston. In a further section on the National Trust in the final part of this study, High Yewdale is taken as a case study of many of the issues surrounding the continuation of traditional fell farming and Herdwick sheep breeding.

Reform of the CAP
The year of FMD, 2001, was also a highly significant year in the further reform of the Common Agricultural Policy. The Hill Livestock Compensatory Allowance was abolished and replaced by the Hill Farm

Allowance. There was a substantial reduction in the budget for the Hill Farm Allowance, which was an area-based payment and which broke the link with production by getting rid of headage-based payments. The context was the implementation in England for the period 2000 to 2006 of the Rural Development Regulation which brought with it budgetary increases for all sorts of non-entitlement based payments, e.g. for organic conversion, for processing and marketing and adding value, diversification and, of course, agri-environment schemes. Such was the concern at the time that a 'Task Force for the Hills' was appointed by the Ministry of Agriculture in late November 2000 with a remit to identify ways in which the Government could help English hill farmers, "to develop sustainable business enterprises that contribute to the upland economy, society and environment, drawing on the full range of England Rural Development Programme schemes and other measures."

Not surprisingly the Task Force picked up much concern about the lack of a focus on the problems of the hills. In the foreword to his report, published in March 2001, the Task Force chairman, (who was the late David Arnold Foster, the chief executive of English Nature) called for, "urgent need for action to sustain the landscape and wildlife which is now as valuable a product as the stock hill farmers traditionally produce." In addition he felt that there was a strong need to educate farmers about a non-headage basis to support what he called a "sustainable future." His Task Force's main recommendation was a basic Hill Environment Land Management (HELM) payment, based on simple criteria and rewarding farmers for production and maintenance of landscape, wildlife and cultural heritage. Regrettably, the Task Force's report with its very pertinent recommendations largely got lost in the FMD crisis of that year.

Another report produced around the same time in 2001 was an English Nature document called *State of Nature: the upland challenge*. English Nature felt that as a consequence of support systems, "many hill areas now hold more sheep than is environmentally sustainable."[26] The report attempted to draw together a large agenda for the less favoured areas of England. It was argued that, "Behind the face of scenic beauty... the English uplands are suffering from economic crisis, social change and environmental degradation." A dire picture was painted of much of the quality of the uplands of England being destroyed largely as a result of over-grazing resulting from hill farmers responding to European and UK incentives to increase their production of sheep meat. The Government agency, however, had a vision for the uplands being, "a mosaic of more diverse habitats supporting characteristic wildlife and at the same time environmentally sustainable agriculture, economies and communities."[27]

In a section on agriculture these points were amplified. It was suggested that there was a crisis in the industry with LFA farmers, "struggling hard to retain viable businesses against the strength of sterling, low market returns for livestock products and the knock-on effects of BSE." This was in spite of the fact that numbers of breeding sheep in the LFAs had risen by around 35% between 1980 and 2000. The result was that, "Many hill areas now hold more sheep than is environmentally sustainable." There had also been quite a strong trend against mixed grazing by cattle as well as sheep in the hills, with a significant proportion of SDA farms giving up their cattle and concentrating just on sheep. Over-grazing was the principal concern in the uplands and the commons were, "in even poorer condition than other upland areas." This was an issue of particular significance for Cumbria, which has 30% of England's common land.[28]

"The way forward" given in the document, although it had the ending of over-grazing at its heart, was impeccably well-balanced. As well as more diverse habitats, environmentally sustainable economies and communities were also required and it was admitted that, "sustainability is about more than just biodiversity." But there did need to be a shift of funding away from production (through the old HLCA and especially the SAP) towards rural development and environmental measures. To solve the problems, four main things were suggested: 1) targeting agri-environment schemes as bio-diversity priorities, 2) reforming the EU sheep meat and beef regimes, 3) enforcing the over-grazing rules more effectively and 4) promoting the implementation of the economic and rural community measures of the emerging Rural Development Programme. In the event, however, very little use was made by English Nature of the available economic and social measures. The interventions were almost entirely about over-grazing as their interventions took place in the aftermath of the FMD epidemic of 2001.

5: The Way Forward for Herdwick Sheep

Recent Issues in Herdwick Sheep Keeping

The Herdwick sheep production system is a demanding and difficult one. Because of the terrain on which the sheep run, fell flocks usually have a lambing percentage of less than 100%, with 85% being common on the harder farms. That is to say, not every ewe rears a lamb, and very few have twins under high fell conditions, though this is increasing where there have been big reductions in numbers due to agri-environment cuts which in turn create new management issues. On the more difficult farms, sheep are not put to the tup until after they have been clipped twice, their first lambs not arriving until the ewe is three years old. Virtually all ewe lambs have to be retained on the farms in order to keep flock numbers up after the inevitable annual draft from the fell of some of the older ewes. The cross-bred lambs on any farm, and the surplus males, represent the saleable lamb crop.

For some time now it has been the case that well over half of the total income of the Herdwick fell farms have come in the form of compensatory payments for producing under 'less favoured' conditions, and for general support for sheep allowed under the terms of the Common Agricultural Policy. These payments, as a result of further CAP reform, have now been de-coupled from production, and the various schemes converted into a 'Single Payment' which over time becomes an entirely area-based payment. There will no longer necessarily be any financial incentive to keep the inevitably lower profitability sheep on the fells. Herdwick breeders may well opt increasingly to keep the minimum (with a margin for safety) number of sheep on their fell ground, necessary to sustain their landlords' flocks or to satisfy their flock replacement policy.

Already Herdwick breeders typically have increased their production of cross-bred (for instance Texel cross) lambs on account of the greater income they generate. As the breeding of pure-bred Herdwicks inevitably declines, a new balance will need to be struck. There will have to be enough fell-going sheep to maintain an adequate heafing system. What we now have in the Lake District is a private sector, community-led grazing system, which often works well. Although no doubt this grazing system will become more geared towards conservation grazing, it is important that it retains a realistic commercial scale.

Grazing for Conservation? Or the costs of Conservation Grazing

Herdwick sheep today carry out a role as animals whose grazing has a conservation function. It is difficult for many fell farmers to make this adjustment in perception as to what they do, but increasingly on account of support payments being tied to grazing prescriptions they are beginning to accept this. Everyone accepts that sheep eat vegetation and it can be the case that the more sheep there are grazing the fells, the worse the vegetative state of the fells. Recent de-stocking has no doubt enabled significant change for the better in vegetative outcomes to occur – and many people are very happy about this. There may well, however, come a time when there will be insufficient grazing on various fells to maintain the desired vegetation and ultimately this might militate against public access to the fells in some areas. There is clearly debate as to what are optimum grazing levels, both in relation to initial 'recovery' and to subsequent 'maintenance.'

It must be pointed out that if grazing falls below adequate levels there will be vegetative changes that on balance will not be regarded as desirable. Nobody really knows what are, for instance, the medium and long term effects of a grazing level of 0.5 ewes to the hectare on, say, the Buttermere Fells. If there are not sufficient numbers of sheep to deliver the public goods of the fells because there are no farmers prepared to run the necessary stocks, to deploy their traditional knowledge and skills, and to take the financial risk, then other mechanisms have to be found. In short, you quickly come up against what might be called the New Forest problem. In 1990 a trio of senior civil servants (from the Forestry Commission, the Ministry of Agriculture and the Department of the Environment) were asked to look at, "the question of the exercise of common rights of grazing and of their importance to the maintenance of the character of the New Forest."

They were asked to look at appropriate levels of grazing to maintain the traditional character of the forest and to recommend ways of getting those levels. They came down overwhelmingly in favour of an option that was based on attempting to regenerate grazing through the use of commoners' animals. They reached this conclusion by looking at other options: grazing by licence, mechanical control and various intensities of introducing a public herd. Although they acknowledged that some people might welcome the direct control a public herd would bring, they doubted its efficacy in a probable situation when common grazing had virtually totally disappeared and all the animals would be new introductions and would not know the "haunts and runs." This would probably lead to fencing having to be erected that would have negative impacts on both access and landscape. "Finally," the report added, "the benefit of commoners' knowledge and experience would be lost." The other great issue about the public herd was cost. The capital cost of establishing a public herd of 3,000 ponies and 1,500 cattle would be about £4 million and the annual cost would be in excess of £350,000.[1]

In recent years Grazing Animals Projects have become very much in vogue especially in lowland England where there has been a steady decline of farmer-led grazing and where conservationists in local authorities, Wildlife Trusts and other amenity bodies have seen the need to introduce or re-introduce grazing as a management tool. This Grazing Animals sector is now quite considerable, with numerous paid posts, a substantial network, the production of guidance and advice on the handling and welfare of grazing animals, the training and deployment of volunteers in tasks like 'lookering' as this network describes the daily visual inspection of stock. There is also a *Grazing Animals Handbook* that categorises different breeds of cattle, sheep and ponies as to their suitability for particular situations.

Although such quango-led and public sector-led and organised grazing has not yet come to the Lake District, there must be a serious presumption about trying to avoid it. There is, after all, a community out there now which has the stock, the skills and the traditional knowledge. That community has both an environmental management role as well as a cultural landscape role – and the major requirement is to sustain it into the future. Officials are, however, examining scenarios and partial abandonment is clearly one of them. It is known also, for instance, that they have calculated what it might cost to re-introduce suckler cows into the hills if there were no farmers, land and buildings had to be found, cattle provided with feed and generally being looked after and managed by skilled and knowledgeable staff. The cost is thought to be £3,000 per cow.

It is frequently asserted by Lake District hill farmers that at some time in the not too distant future there will be a policy-led drive to re-build the numbers of fell-going sheep. It is not uncommon for farmers to say something on the lines of, "they will be paying us to put them back on in ten years time." But this is usually supplemented with a statement that this should not be taken for granted especially if, in the meantime, there is further erosion of the numbers of people with the necessary fell sheep management skills and knowledge. There needs to be a continuing transmission of these skills from one generation to the next.

Farm Animal Genetic Resources

Another concern revealed by the FMD crisis was the threat that epidemics amongst sheep posed to breeds which are very highly geographically concentrated. The Herdwick is found almost in its entirety in the English Lake District and there are a number of other breeds in the same situation, such as the Rough Fell with its high concentration around Kendal, Sedbergh and Tebay, and the South Country Cheviot with its concentration in part of the Borders. During FMD, Professor Dianna Bowles (an eminent biological scientist from York University, herself a breeder of Herdwick sheep) established the Sheep Trust and secured resources to collect and store genetic material from breeds, including the Herdwick, which were threatened either by disease or disease management. The Sheep Trust developed a powerful case for the acceptance of a category of livestock breeds as 'Heritage' breeds, where there was a genetic resources case which also often had a cultural landscape value.

The publication of the UK National Action Plan on Farm Animal Genetic Resources in November 2006 saw this case being made quite precisely with Herdwick sheep being used as the outstanding example. The FAnGR report pointed out that during FMD several breeds suffered 'huge losses.' The Herdwick, Rough Fell, Cheviot and the British Milk Sheep all lost more than one third of their total population. It reported that about 30,000 Herdwicks died during the FMD outbreak, "including almost the entire generation of hoggs wintering away from the home farms." Developing the argument, it went on to say that:

"The Herdwick, in particular, illustrates the value of 'heritage' breeds. The local adaptation of its flocks to the Cumbria fells offers environmental and marketing benefits. The breed is intimately linked with the Lake District, providing a strong marketing base through this heritage."[2]

The underlying reasons why the UK has produced this Action Plan stem partly from the fact that the UK signed the Convention on Biological Diversity at Rio de Janeiro in 1992, and partly from more recent concerns about the global spread of relatively few specialised breeds and the consequent loss of genetic material that in changed (for instance, climatic) circumstances might come into its own. Perhaps the best example from the UK of this occurring relates to the Lleyn. Although its rise has nothing to do with climatic factors, the Lleyn was virtually extinct as a commercial sheep breed until a generation ago when it was rescued from decline in its native area of north west Wales. It is now an extremely popular and adaptable breeding sheep which in some areas, including Cumbria, is challenging the Mule.

There are about 120 farms which keep some Herdwick sheep in and around the Lake District, but perhaps 100 which keep Herdwick sheep in significant numbers on the fells. It may well, therefore, soon be the case that the number of recognisably pure-bred Herdwick sheep carrying out a traditional environmental grazing role in the Lake District will fall below the European Union level that categorises a breed as being "in danger of extinction." This level is defined at 10,000 breeding ewes of sufficient quality to be registered in a flock book. Farmers keeping these threatened breeds are eligible to receive support payments on a headage basis. Many of the commercial Herdwick fell sheep would scarcely be fit for registration in a flock book as good examples of the breed type, so it might well be possible that the Herdwick might qualify.

Although this provision is used by every EU Member State except the UK and Denmark, it is apparent that Defra does not propose to use it here largely it seems because it is based, (as it perfectly sensibly has to be), on a headage payment basis.[3] If, however, Defra really wanted a mechanism to retain native hill breeds such as the Herdwick, it could readily be done through a managed and regulated process, based on something as simple as inspection of the annual production and retention in the flocks of gimmer hoggs (ewe lambs) probably using a mechanism quite similar to the old HLCA regulations. It can, therefore, only be concluded, regrettably, that Defra adheres on this matter to a dogma about the non-use of headage payments.

A recent decision on Agricultural Biodiversity, to which the UK signed up, included a call for signatories to mainstream agricultural biodiversity in their plans, "with the active participation of local and indigenous communities" and to recognise and support their efforts in conserving that biodiversity.[4] All this is, indeed, welcome news since it implies that there is some national priority (subject to the existence of a joined up approach) being given to what might be called the *in situ* conservation of Herdwick sheep, that is to say their continued existence as a grazing animal in the Lake District providing both marketable products and public goods.

Adding Value to Herdwick Products: An Agenda for the Breeders

Can Herdwick farmers do anything for themselves to improve their incomes? One obvious answer is to attempt to add value to Herdwick products. Although, as Professor Philip Lowe puts it, it is "readily apparent" that the public benefits of pastoral farming in places such as the Lake District, "far over-shadow the market value of its tradeable products,"[5] there is potential for marketing some high quality local products in an added value form and where the connection to the landscape is made explicit. Some members of the Herdwick community are trying to add value to their own products and to 'cash in' on the positive connections that clearly exist between the Herdwick, its working landscape and living heritage. In this respect at least some members of the Herdwick community have comprehended the sustainable agriculture message of the Curry Report on 'Farming and Food' of reconnection with consumers and adding value to their product.

Herdwick meat has long had a very high reputation for its eating quality. It was eaten at the Coronation Dinner in 1953 and has always been the favourite amongst local farmers and butchers for their own consumption. In 1997 a scientific study of meat quality was carried out by the Food Animal Science division of Bristol University's Veterinary Science School which established that the reputation of Herdwick was, indeed, well founded. Herdwicks of eight months of age and Herdwicks of twenty months of age were investigated for their eating quality with cross-bred Suffolk lambs of six months of age being used as a control. Overall, it was concluded that, "the results provided evidence for high levels of eating quality of Herdwick sheep." It was also noteworthy that although the older Herdwicks were thought to be tougher, this did not lower the overall liking score as this toughness was counterbalanced by "superior flavours."[6]

Some producers have started in the last few years to realise the potential of the meat. For instance, the pioneering activities of Eric and Susan Taylforth of Millbeck, Great Langdale, and the enterprise of the Relph family at Yew Tree, Rosthwaite, who won the *Farming Today* national prize for 2002 at the BBC Radio 4 Food Programme's awards.

Expert butcher Andrew Sharp, trading as Farmer Sharp, has also secured national recognition particularly amongst leading chefs and in 'foodie' circles with his weekly sales of Herdwick meat at Borough Market in London, and his efforts to promote Herdwick meat. He is involved in the UK section of the Slow Food Movement and participates in the prestigious Italian Food Festival, Salone del Gusto (supported by the Fells and Dales LEADER + Programme). He buys his Herdwicks from a supplier group largely in the Langdale area. Various other breeders have developed or are developing their direct marketing efforts – though there is still a long way to go before every Herdwick lamb or shearling is sold at a premium as Herdwick meat and not just submerged in the anonymity of the food chain.

LEADER grants have assisted several farm-based meat schemes on National Trust tenanted farms, notably at Yew Tree, Rosthwaite, Yew Tree, Coniston, and also at Wasdale Head Hall. Some traditional local butchers have Herdwick lamb available and have strong links with their supplier farms. Quality food outlets are also beginning to see the commercial advantage in having Herdwick lamb on sale: Plumgarths near Kendal, and Cranstons Food Hall in Penrith, being prominent examples. Enterprising breeders have developed relationships with quality local butchers and are supplying them with considerable numbers of lambs, with Anthony Hartley of Seathwaite being a leading example.

A major breakthrough came in 2006 as a result of work being carried out by Veronica Waller of the Fells and Dales LEADER + Programme on creating a 'Lakeland Herdwick' identity and a direct sales market. Aspects of the development work were funded by the Sustainable Development Fund of the Lake District National Park and a grant from Friends of the Lake District working alongside the Herdwick Sheep Breeders' Association, with the National Trust as a supporting partner. As part of this work a box scheme, Lakeland Herdwick Direct, was developed and ran moderately successfully for a couple of years although it failed to make a breakthrough into sufficient numbers to sustain itself as a scheme.

The concept, however, of linking the meat with the landscape qualities of the area of production, became well established and was taken forward in 2006 with an arrangement made between a small group of Herdwick producers (mainly from the south west of the Lake District) and Booth's, a good quality regional supermarket operating throughout the north of England. In 2008 the scheme took 78 lambs per week over an approximate five month season. These are sold in their twenty or so stores which have a fresh meat counter. The farmers receive a premium, their meat sells well, and the point of sale material indicates the provenance of the lamb, linking it to the Herdwick story.

There is obvious potential for Herdwick to be branded: it is a highly specific, locally adapted and traditional breed which comes from a very special landscape full of associations with high landscape quality and environmental value. The associations of the breed with the Lake District and its links with the National Trust reinforce these attributes, as the Trust acknowledged after the foot and mouth outbreak with the appointment of a Herdwick promotion officer, a post occupied until early 2007 by David Townsend. There is also potential for Herdwick meat to have protection preventing other meats being passed of as Herdwick by seeking European Union protected food name status, through either a PDO (Product of Designated Origin) or a PGI (Protected Geographical Indication). 'Lakeland Herdwick' might come to have the same resonance as a distinctive local food such as Parma Ham, the food product around which the protected food names system was originally designed. A case is being developed for submission to Food from Britain with the hope that Defra will back it and take it to Brussels.

Since the mid-1990s, the market has been saying remarkably consistently that Herdwick wool is surplus to requirements. There is a world over-supply of wool, Herdwick is a coloured wool (which is less versatile than white wool), there are currency conversion problems and there is the inexorable rise of synthetic fibre: elements of a familiar story. Meanwhile the sheep need clipping and fleeces need to be disposed of. The British Wool Marketing Scheme of 1950 requires every owner of more than four sheep to register with the Board and to send in their wool at the price laid down in the annual price schedule. For the first few years of the Wool Board things were relatively favourable. Up until the early 1960s a relatively large proportion of hill farm income derived from wool – on some hill farms as much as one third.[7] Wool was a significant part of farm income even though the price of wool fluctuated wildly. For instance, the late Scott Naylor told me that when he took over the tenancy at Middle Row, Wasdale Head, in the early 1960s he had a wool cheque of £600 which was four times the then rent and a man's wage was £15 per week.

On some occasions Herdwick wool has earned relatively high prices. There was one year in the 1980s (when the Japanese entered the market in a big way) when it achieved over £1 per kilogram. A happy calculation for a number of years in the 1980s was that the wool cheque averaged about a £1 a fleece.[8] In the meantime the guaranteed price system that used to be in place was abolished, and each grade of wool had to depend on its own market price. Herdwick wool, in its two shades of light and dark, has been for a number of years now worth just a handful of pence per kilo.

Although three or four years ago when this desperately weak price for Herdwick wool began, farmers felt very uncomfortable about burning their wool, many of them are now getting used to it. Wool burns well straight off the sheep on the day of clipping; there are no fleeces to wrap; no wool sheets to pack and to lift and to store – and no bills to pay to transport the wool to the warehouse. A good and interesting commodity has essentially become a waste product, prompting a disposal rather than a marketing solution. The Wool Board legislation designed to correct market failure has itself failed with regard to low value wools. A system, designed by the state (and retained by the state when all the other marketing boards have been closed down) to intervene on behalf of the producer in an uncertain market does not correct market failure: rather it stifles innovation and generates environmental costs.

The Wool Board is very aware of the Herdwick wool price situation and spends some time trying to interest wool users in Herdwick wool and to work on new product development. These efforts are clearly based on a belief in the 'trickle down' theory on which the Board works: the greater the amount of products made from Herdwick wool, the greater will be the demand in the wool auctions for Herdwick. Regrettably however, this has yet to work in spite of all the efforts that have been made. Within the last ten or so years, for example, there have been three new Herdwick carpets put on to the market (admittedly of greatly differing quality): all three have been launched to all the right noises invoking the fells, the Lake District, the National Trust, with Beatrix Potter getting her obligatory mention. Given that carpets have more potential than most other products to consume high volumes of wool, this should have worked – but the fact remains that Herdwick stays at the bottom of the pile – and now the issue is not so much one of demand but more one of supply.

The most recent carpet, a good quality Axminster, produced by a Cumbrian firm in the historical centre of the Cumbrian wool trade, Kendal, may find it difficult to source wool unless the producers of the wool can be given sufficient incentive to send it in to the Board: a price of at least 50p per kilogram is widely seen to be the lowest acceptable price and nobody currently is expecting that the Wool Board's price for Herdwick wool will reach these dizzy heights. Thanks to a partnership deal between Goodacre's of Kendal and the National Trust, a small subvention was made available to support the wool price on the 2004 clip. Just over 100 Herdwick wool producers have enjoyed a small level of support which might have, at most, covered the clipping costs.[9] Not that there is any money to be made at this price but at least there is no loss given that the sheep have to be clipped anyway for health and welfare reasons. Herdwick breeders as a group are proud to see their product turned into something useful and/or beautiful: but they have wondered over the past few years as to why they, the primary producers, fail to benefit whilst everyone else involved in the fibre chain seems to.

Although it is anticipated that this level of support to a quality local product will remain in place for a further two years, there was disappointing news in November 2005 when Goodacre's announced that they were transferring all their carpet manufacture, including the Herdwick-based range, to Poland where labour costs were very much cheaper than those in Kendal.[10] Not long after there were further changes with the wool itself being sent to Turkey to be spun and Goodacre's going into receivership. The whole story, regrettably, has been a triumph for globalisation rather than for the local produce economy which is where the story began.

In addition to the Herdwick carpet initiative, some farmers are also trying to add value to Herdwick and other fell wools, though it must be pointed out that they do so against the background of the statutory instrument which established the British Wool Marketing Board which seems to stifle rather than to encourage enterprise and innovation. The Herdwick Sheep Breeders' Association gave written evidence in 2003 to the Select Committee on Environment, Food and Rural Affairs suggesting that, without dismantling the arrangements for the Wool Marketing Board to do what it currently does; those wool producers who wanted should also have the ability to add value to their own product. The evidence concluded by saying that, "Amongst the producer community there is a wealth of energy and initiative which can most usefully be set free by allowing producers and groups of producers to have the option of adding value to their own product."[11]

New Markets for Live Sheep?

In spite of various pious hopes, the Herdwick's range as a commercial sheep has not extended much beyond the Lake District.[12] True, every now and again commercial hill farmers from Derbyshire or Dartmoor come to the back-end sales and buy substantial numbers of ewes and some come regularly to purchase tups. But there is a growing number of small breeders from all parts of the UK who are now keeping small flocks of Herdwick sheep. As early as 1985 the HSBA Executive considered the issue of the large number of small breeders who were seeking to join the Association and what could be done for them.

This was really the first glimmer of what might be called the 'Beatrix Potter Effect,' whereby smallholders from all over the UK (and sometimes further afield) want to keep Herdwick sheep not particularly as commercial propositions but as animals they found interesting and attractive. Some Lakeland breeders have developed small markets to satisfy part of the demand for private sales. Prospective breeders have sometimes come to the Lake District to buy stock at sales, but generally they have purchased from the growing numbers of sheep of an approximate Herdwick type that can be found in auctions up and down the country, often incorrectly sold as specimens of a 'rare breed.'

For some years the Association gave sanction to there being an official sale of rams at the National Sheep Association's Wales and the Border Ram Sale held annually at Builth Wells. This enabled small numbers of Herdwick female sheep also to be offered for sale. In order to ensure some quality control the Association appointed an inspector of these sheep, but after several years the experiment was brought to an end as a result of the quality not being good enough. In addition, various officers of the Association have visited farms up and down the country to see sheep and occasionally to register tups.

Effectively, however, the best tups clearly come from the Lake District – and there are plenty of them. Over the past few years at least 350 tups are registered annually by about 60 breeders – although about ten farms and breeders contribute about half the number. For example, the Hartleys at Turner Hall register about fifteen to twenty tups a year as do the separate Richardson families at Watendlath and at Gatesgarth. The Blands at West Head, Thirlmere, have in recent years produced in excess of 40 quality tups annually. The very small flocks in other parts of the country create virtually no significant demand for tups – by my estimation perhaps only 20 or 30 a year are sold to be used outside the Lake District. The flocks outside the Lake District are certainly not making up for the reduction in Herdwick ewe numbers inside the Lake District.[13]

In October 1995 the Association attempted to develop an organised way to meet the demand from outside the Lake District for small numbers of female sheep and crucially to give people a simple way of accessing a substantial number of good Herdwicks. A special sale was held at Penrith Auction Mart, chosen for its accessibility from the M6. Just fewer than 200 females, sold in small lots, and 42 rams, sold individually, were presented for sale. The sale attracted some outside interest, but it was very much the case that the sheep of quality presented there essentially went to Lake District buyers and so it was decided not to repeat the experiment. Since then the Association has remained concerned as to how to encourage interest in the breed outside the Lake District and to balance this against the resources of time and money that might be required to visit small numbers of sheep in many locations for an end that might not be very productive of the Association's essential mandate of keeping vibrant the breeding of Herdwick sheep which graze the Lakeland fells.[14]

The Association does, however, respect and wish to support the enthusiasm of Herdwick breeders from outside Lakeland – not least because of the potential gene pool that might exist outside Lakeland in the case of another disease epidemic such as that of 2001. In 2007 a group of enterprising breeders supported a new sale at Bentham Auction whereby people could sell individual Herdwick female sheep. The sale in its second year attracted an entry of 46 sheep from six breeders. The Association has also discussed the possibility of holding some 'master classes' in Herdwick sheep breeding and a suggestion has been put forward of allowing the formation of a Herdwick

Breeders' Club for enthusiasts in the rest of England and Wales – and Scotland too if breeders there required it.[15] A development of this sort need not and should not detract from the Association's key role of promoting and maintaining Herdwick sheep breeding in the cultural landscape of the Lake District.

6: Herdwick Country

Herdwick Country: A Cultural Landscape

Herdwick sheep are a key component of the cultural landscape of the Lake District. Keeping Herdwick sheep on the fells is as important as keeping the lakes and becks clean, the woodlands managed, the stone walls maintained and the vernacular farm buildings repaired. Much of this cherished landscape has been created round the keeping of sheep. The pattern of the intensively grazed fields of the valley bottoms; the walled intakes on the lower slopes; and the large expanses of rock strewn open fell above the fell walls make up the typical landscape of the central area of the Lake District National Park. Herdwick sheep keeping has made and continues to make a vital contribution to the management and maintenance of that landscape. In Herdwick Country you can see a landscape of great antiquity shaped by the keeping over hundreds of years of sheep and cattle. The very lie of the land has been altered to utilise the grazing resources of the area – for example, stones have been cleared from fields and ways have been created to get sheep to and from the fells in the form of 'outgangs' and 'ingangs' and conventions have been established about how to manage communal grazing on large expanses of fell.

When these physical and cultural systems were first established hundreds of years ago, they were the concern only of the community of graziers and their landlords, and subject to regulation and policing through the manorial courts. Now there are numerous stakeholders and organisations with an interest in these things as a result of the development, largely in the second half of the twentieth century, of concerns about landscape, environmental protection and the promotion of public access. The National Park Authority (with its statutory duties) and the National Trust (as the largest land owner) are, of course, the two bodies, alongside the fell farming community itself, which carry the main responsibility for the continuing management of the Lake District. The text in previous sections has dealt with the activities of all three groups over the years in terms of managing and maintaining the essential resources against all the challenges of the operation of a free market in property – which all the time threatens to undermine the management system which has both created and sustained a landscape which is truly of outstanding universal value. Can these efforts be adequately sustained? The next two sections look at the National Trust and the National Park Authority in the light of this question.

The Role of the National Trust and the Lessons of High Yewdale

The National Trust is on record as saying that it recognises that, "one of the most effective methods of managing the countryside is through the practices of farmers."[1]

One way of establishing some basic information about what has been happening to the National Trust's Lake District estate can be found by analysing the two contributions on 'The National Trust and Hill Farming' in the versions of the *Lakeland Shepherd's Guide* published in 1985 and 2005 respectively. In 1985 the Trust owned 82 farms, "bought or given to save Lake District valleys from destructive development." These farms are listed with 54 of them having a landlord's flock, "mainly of Herdwick sheep" of which the Trust has a total holding of 22,000. Twenty years later the Trust owned 92 farms, 58 of which had landlord's flocks. The total landlord's stock, "mainly of Herdwick sheep" is still given as 22,000.

One farm has disappeared from the list over the period. This is the small farm, Kidbeck, at Nether Wasdale which, after the retirement of a long-standing tenant, was re-let without a landlord's flock heafed to the common, it having been dispersed by the Trust. Some additional fell farms based on Herdwick sheep breeding have also been acquired, notably Fenwick, Thwaites and Bowderdale, Wasdale.

There has been a substantial turnover of tenants in recent years, some of which was inter-generational under the old legislation: farms concerned here probably being High Birk How in Little Langdale, Mireside in Ennerdale, Fold Head at Watendlath, Brimmer Head at Grasmere and Seathwaite in Borrowdale. Over the past ten or so years to my reckoning, however, there are about nineteen farms (all with landlords' flocks) which will have been let on Farm Business Tenancies. Within this number twelve farms have been let to people from Cumbria and five to people from outside Cumbria, but generally speaking from other hill farming areas including County Durham, Yorkshire, Derbyshire and Dumfries and Galloway.

The final two farms making up the nineteen, Wha House and Kidbeck, have been let twice in recent years, the first time, unsuccessfully, to people from outside the area and the second time to local people. Included amongst the new tenants are people with excellent stock management skills and amongst the number are some enterprising individuals and couples. The key issue is not one of locals versus non-locals, but rather the importance of people coming to the Herdwick farms respecting local practices and knowledge, and integrating with the local communal effort whilst necessarily pursuing their own agenda for viability.

In the *Shepherd's Guide* article it is stated that, "The Trust intends where appropriate to maintain its involvement in sheep flocks in the foreseeable future." It gives its objectives as being to protect the beauty of the landscape for the benefit of the nation and to help perpetuate the valley communities. It indicates that in practical terms farms are "modernised" without compromising the traditional look of farm buildings. Some indication is also given that the Trust is actively involved in campaigning for improved farm incomes.

Although everyone in the fell farming community understands the importance of viability, it is widely realised that some of the smaller farms struggle to be viable on their own account, and that it is only the enterprise – both on and off the farm – of the families that occupy them that enables adequate income to be gathered. This is well recognised and accepted and is the case on the non-National Trust farms also whether rented or privately owned. It was a huge surprise, therefore, early in 2005 when it was announced that High Yewdale at Coniston, was going to be split up when its tenants John and Ruth Birkett retired later that year on the grounds that the farm would not in the future be viable, even though it had been run successfully for several decades without diversification into tourism or off-farm working. The proposals for Yewdale raised serious concerns about the future of the NT estate.

High Yewdale was regarded as one of the best of the National Trust's Herdwick farms. It was relatively well structured in terms of its in-bye resources compared to its fell ground; its fell flock was well heafed and it was exceptionally well-managed. Yewdale had been used as a showcase for the Trust's work, for instance by hosting a visit from the Queen. The Birketts had, like many long-standing tenants, acquired land of their own which they worked in with the farm. The Trust calculated that Yewdale was a farm which would see a sharply reducing Single Payment and projected that a new starter would not have additional land available to them – although its more appropriate letting to an established small tenant farmer wishing to progress might not have presented this problem.

The initial decision was made largely it seems because the local Property Manager had no budget for the £120,000 or so required to provide a modern cattle building at Yewdale. A decision, therefore, was made to end the existence of Yewdale as a farm and to offer parts of its lower ground to neighbouring Trust farms (which like many other farms could do with strengthening) and to put the fell, the heafed sheep and some of the inbye to neighbouring Tilberthwaite. However, rather than this being a welcome development the Wilkinsons at Tilberthwaite regarded it as a complete betrayal of the National Trust's stewardship. They robustly rejected the offer saying that they wanted a neighbour not 40 acres of inbye, 600 sheep and another fell. The other farms, however, did not demur from having additional land to strengthen their holdings and the fell, the fell sheep and some of the inbye land were eventually put to Boon Crag.

A huge controversy raged in the Lake District, with many letters of opposition being published in the *Westmorland Gazette* from the farming community, local people and organisations, and concerned individuals many of who saw the proposed arrangements as a betrayal of Beatrix Potter's original motive in acquiring farms – against which the Trust posited her hard-headed business acumen.[2] But a process had been set in train whereby the Trust committed itself to a particular road, with any divergence from that being regarded as a vote of confidence in its stewardship of its Lake District estate. The Trust backed its decision with a large number of arguments, trying to din into an uncomprehending population what they regarded as the realities of hill farm economics, and the future picture of traditional farming in the Lake District. There were all sorts of justification: the farmer was 'iconic' but not the farm; the farm next door (Yew Tree) was more 'iconic'; it was anyway all economics really and the best way to support three other local farms which were all struggling; the heafed fell sheep would go to a new home.

Peter Nixon, the Trust's national land agent, in April 2005 stated that the justification of the High Yewdale decision was that "the harsh truth" was that sheep farming in the Lake District without subsidy was a "loss maker." He argued that the future net farm income at Yewdale would not provide a reasonable living for a tenant, and that the Trust had a moral responsibility to avoid creating a situation where a tenant was bound to fail financially and possibly thereby lose their life savings. If a tenant were of necessity required to be part-time there would be a "high level of reliance" on "non-Trust related, off-site income." The Trust would lack control over this. He also pointed out that the existing tenants had not carried out any diversification activities and that it would therefore cost a lot of money to convert buildings for these purposes. He continued:

"occupying High Yewdale as an independent tenancy with the associated land maintenance as well as stock management activities would be a full-time occupation for a tenant so allowing very little time for other income generation activity; and the risk that any tenant would become compelled to squeeze every ounce of productivity out of the holding, risking compromising the Trust's conservation objectives in the process."

But, fell farmers asked, could many of the farms withstand this type of analysis? Within this argument there was effectively a concession that the viability of the Trust's farms was poor and therefore that the usual way of managing them as conservation assets via conventional tenancies (whether farm business tenancies or traditional tenancies) might not any longer be tenable. Little wonder, therefore, that some discussion emerged about the concept of Trust tenants receiving a wage (in the form perhaps of a rent reduction or even a negative rent) as part recompense for the work they did.

The chairman of the Open Spaces Society echoed the Trust's position about Beatrix Potter's business sense and urged the board of trustees to resist attempts to turn the Trust into what he called the 'National Farmers' Preservation Society.' He said, "I can't see what all the fuss is about. The farm buildings are still there. The fields are still there. The hedges are still there. The walls are still there. The sheep are still there." He did not think that the Trust should start "subsidising its farmers." He went on, "Some want to see the Trust do just that. Instead of collecting rent we shall be paying farmers to live in the Lake District."[3]

The fact of the matter, of course, is that Yewdale is no longer there as a whole farm and it fundamentally raised questions as to how the National Trust intends to maintain its fell farming estate. How, in the future, will the National Trust protect and maintain the heritage assets it has acquired? Is the business model, effectively, one which might be characterised as "each farm must make a proper full-time living for a farming family"? If this is the case the farm amalgamation process might be difficult to stop. Or is it one where new ideas and a different business model are brought to bear?

There could be a vision for the future of some of the smaller Herdwick farms whereby the people farming them earned their living partly through their enterprise as a farmer, and partly from doing a job for the National Trust for managing, on contract, the land which they farmed with this job being based on a realistically achievable work load and proper payment. One could imagine that the tenant at Burnthwaite at Wasdale Head, for instance, might on that basis, be paid the going day rate for the 50 days a year he reckons to put into the maintenance of the historic wall system on his farm. Currently this is part of his obligations as a tenant, incidental to his management of the sheep on the occasions when they are not at the fell. It might, in future, be part of his income earning portfolio, along with his tradable products of calves, lambs, wethers and ewes, his agri-environment income, his Single Payment and the income from his family's tourism and hospitality enterprise.

This would, of course, upset some people. The National Trust is an environmental conservation charity and the Herdwick farms it owns are presumably in its hands because, over the years, it was felt that they represented a key part of the cultural landscape and heritage of the area. Up to now they have largely been managed through conventional farm tenancies. Now, however, with the Trust calculating that subsidy income on farms will fall by 40% by 2012, the prospects on the farms are so poor that the Trust reportedly finds it difficult to "find suitable people to take on the farms." These are the words of Neil Johnson, one of the Trust's farming and countryside advisers and personally committed to fell farming, speaking at a Tenant Farmers' Association meeting in Carlisle in February 2007.

Although there was a shortage of prospective tenants with the appropriate skills and knowledge, Johnson stated that there were numbers of what he called "lifestyle" applicants interested in the farms as attractive places to live and no doubt bolstered by other sources of income from outside farming. Running with this development (as Johnson is aware) would be a matter for regret, given the importance of traditional skills and knowledge in managing these farms. George Dunn, the chief executive of the Tenant Farmers' Association and a member of the Trust's land use and recreation panel, suggested that, although the rents on the farms were low, it was very difficult to make a profit on them. Alongside local branding and small-scale co-operative ventures, "we are having to look," he said, "at new ways of running these businesses. In the future, we might see people paid to farm the type of holding the National Trust owns."[4]

Just as the National Trust employs gardeners to look after historic gardens, so it can employ farmers to look after historic farms and landscapes. This process may, in fact, already have started in other situations. The National Trust on 17 November 2006 advertised in the *Farmer's Weekly* for a 'Farmer' at a salary of £18,310 per year at Llanerchareon, near Aberaeron, Ceredigion, in Wales. The farm was stated to be a popular visitor attraction with a John Nash house, a working organic farm, two restored walled gardens producing fruit and herbs. It was written up as being an in-hand farming venture, "where farming interacts with visitors to demonstrate how agriculture and food production are as one, Llanerchaeron is home to pedigree Welsh Black cattle, Llanwenog Sheep and Welsh Pigs." The duties of the salaried farmer were the day to day management of stock, stock buildings and in-hand agricultural land. As well as running the livestock breeding programmes a key role also was to maximise the income potential of the farm and the presentation of its activities to visitors and school groups.

It would be surprisingly easy to create an equivalent job description for many of the National Trust's Herdwick and traditional fell farms. One could consider, for example, the contributions made by the tenant farmer at Nook farm, Rosthwaite, Borrowdale, which was acquired by the National Trust in 1947. Nook is part of the National Trust landholding in one of the most scenic, culturally landscape rich and bio-diverse dales of the Lake District. The farmer has to manage two traditionally heafed stocks of highest quality Herdwick sheep carrying out essential grazing management of the central massif of the Lake District; this also involves collaborative work with neighbours and an appreciation of common land management issues. The farmer maintains a small herd of beef cattle to carry out valuable pasture management on the farm's inbye and intake ground. The farmer has to take responsibility for the maintenance of the farm's countryside features including very large lengths of stone walls and vernacular buildings. The farmer must farm with woodland management and water resource protection in mind. The farmer must ensure that his farming operations comply with the requirements of the Lake District ESA and any successor agri-environment agreement. The farmer uses low input systems and farms as closely to organic standards as is possible on a common land system. Nook farm has one of the best Herdwick flocks in the Lake District and the farmer is responsible for producing Herdwick tups of high quality for the maintenance of varied bloodlines in the breed.

The farmer contributes to the production of finished Herdwick lamb at high specification suitable for marketing under a potential European Protected Food Name scheme. The farmer is actively involved with a communication programme with the general public through visit and farm walks. The farmer is involved in maintaining community strengths and activities through his involvement in the Borrowdale show and shepherds' meet. The farmer has been involved in the training and mentoring of a young person interested in fell farming. The farmer makes a strong contribution to the critical mass of hill farming activity and to the life of the valley and the traditional fell farming area generally through his involvement as a member of the Executive Committee of the Herdwick Sheep Breeders' Association. A similar conspectus could be produced for many other fell farmers and these contributions to the social, economic and cultural fabric of the area need to be properly valued and rewarded.

Innovative solutions are called for to retain the Herdwick estate. An end needs to be put to what looks like a war of attrition on the small and medium sized fell farms or else, before long, there will only be very large, essentially ranched, farms left. There needs to be a critical mass of farms to deliver cultural landscape management just as there needs to be investment in the human resources for the future. Something like the fell farming traineeships scheme run by Voluntary Action Cumbria as part of the Fells and Dales LEADER + Programme probably needs doing every few years. There also needs to be support to pieces of action research like that being carried out through the Carnegie UK Trust action research project to explore the long term sustainability from the community dimension. It would also be feasible to develop Community Supported Agriculture solutions where members of the public could be invited on an organised basis to buy the produce of farms they 'adopted'.

A definite end needs to be put to the process of farm amalgamations within the National Trust Lake District estate. There needs to be a network of key Herdwick farms, ranging from the dalehead farms through the medium size farms in the length of the valleys. There also needs to be a sprinkling of smaller units which can provide housing and an enterprise opportunity for young people working in the hill farming sector. These small farms seem to be especially prone to being amalgamated into larger holdings because there will probably be no entitlement to receive the Single Farm Payment and the agri-environment income will inevitably be modest as will be the income from tradable products. The only reason they might be working as viable farms today is because the existing tenants on them have developed 'composite' holdings with additional land away from the farm. But the basic small units could more often provide key workers (and they are out there) with accommodation for themselves and their sheepdogs along with their necessary equipment.

It is only a few years ago, in September 2004, that the Trust seemed to understand this. The Trust's then farming adviser pointed out that on many of the small farms it was necessary for income to be sourced away from the farm by the small farmer. John Metcalfe said that, "While there may be enough time to do the essential stock tasks on a part-time basis it can be difficult for some tenants to keep up with general maintenance," but it can be made to work. Stonethwaite Farm, Troutbeck, was advertised in 2003 as a part-time farm and the tenancy given to a tenant who worked about three days a week on his holding managing his 300 ewes and ten suckler cows and the rest of the time he worked as a dry-stone waller.[5]

More recently, however, the fell farming community has cited a case where such sensible arrangements do not seem to have prevailed. The small farm of Milkingstead, in Eskdale, with 50 acres of ground was recently left to the National Trust. The ground was put to two other farms and the house was advertised to all comers to let at offers of rent over £400 per month – a level which was regarded locally as being unaffordable. There was a further stipulation of "no dogs"! Needless to say there were several young people who do farm work in the area who would have welcomed the opportunity to have been properly housed and having some land, whilst still working for farmers in the wider area. There also needs to be encouragement given to farmers to increase the local affordable housing stock by creating suitably regulated key worker housing from redundant farm buildings. Solutions of this sort will need to be found valley by valley if the management of the land which everyone wants to see is to be maintained.

It is heartening to recognise that the mistake of High Yewdale seems to have been recognised – evidenced perhaps by the fact that a small farm like High Snab, Newlands, was in 2008, after a retirement, let as a farm. It is good to hear also that new tenants are being well-supported by programmes of investment on their farms (for example, new sheep pens) and in their farmhouses (for example, en suite bathrooms in the letting bedrooms), even though there are the inevitable gripes about the time and bureaucracy involved.

There is a great deal of good practice on the National Trust estate but great care will need to be taken over the next few years. Every encouragement needs to be given to young people who are interested in hill farming and the keeping of fell sheep. If they have no prospects of living and working locally, they will inevitably drift away from the fells. There even needs to be renewed attention given to the purchase of key fell farms and of better land that could be put to the smaller fell farm units. In short there needs to be a clear return to the historic mission of the National Trust throughout the whole Lake District. It will have to be accepted that new ways will have to be found to sustain traditional fell farming and that this will involve a wide range of partners, not least the Lake District National Park Authority.

An Agenda for the National Park
It has often been suggested that hill farming is in a state of permanent decline and that nothing in the end works well enough to ensure the long term

sustainability of the system. For example, two academics, Davidson and Wibberley, in 1977 described the uplands as consisting of "interlocking vicious circles of natural, economic and social difficulties which act and interact upon each other in ways which are hard to change with any degree of permanence."[6] The Lake District National Park Authority has not only been amongst those organisations which realise the difficulties which fell farming has, but also has made some pioneering and important attempts to ameliorate some of the problems. In 1969, for instance, the Countryside Commission and the Ministry of Agriculture launched an Upland Management Experiment in a small number of Lake District parishes. The experiment gave financial encouragement to farmers to carry out small schemes which improved the appearance of the local landscape and enhanced the recreational opportunities in the area. This was sufficiently successful to be extended to cover a larger area.

From 1976 'Upland Management' became part of the work of the Lake District Special Planning Board (as it was then called) and had a budget in the mid 1980s of £184,000. At one point the Planning Board made a point of employing farmers on a part time basis, but eventually upland management came to be carried out by full-time employees. Although upland management was popular with farmers it was concluded in 1976 by the Countryside Commission that the scheme was only, "scratching the surface of the real social and economic problems faced by the farming community." A similar conclusion was also reached in a special study of Hartsop carried out by Gerald Wibberley in the same year. Upland management was unequal to the task of stemming the tide of the undesirable social and economic changes that were occurring in the area. Bolder initiatives were needed, requiring government intervention to carry out, "measures aimed at attacking rural housing shortages, assisting hill farmers and maintaining a national asset."

The Planning Board (which had amongst its members John Dunning – who had detailed knowledge of hill farming as well as of enterprise development – as a prime mover) in 1980 began a third experiment to try to bend policy at the local level to meet the real needs. It wished through UMEX 3 to integrate tourist development, landscape and nature conservation on 25 farms using Ministry of Agriculture grant-aided development plans. However, there was a mismatch between national policy which was all about the expansion of production and what might now be called a 'sustainable development' approach – the need for which had been effectively discovered twenty years ahead of its time. Consequently no progress was made and the scheme faded away.[7]

Since then the National Park Authority has had something of a defeatist attitude about the possibility of sustaining a hill farming system of the right sort. Analysis of the problems continues apace, but it is increasingly less common to find suggestions as to what might be done. For example, the National Park Management Plan, produced in 2003, pointed out that,

"Farming and the way the land is managed have done much to create the special qualities of the Lake District." These special qualities were distinctive high quality produce, the character of the cultural landscape and its bio-diversity, farm woodlands and the open nature of the fells. Farming, though, had been under, "immense and continuing pressure" and was now in "great difficulty." Farm amalgamation was continuing apace and farms and farmhouses were being sold off to people, "who do not always contribute as much to the local economy or community life, and are less active in managing the landscape."[8]

A similar analysis of the situation is set out in the 2004 Management Plan which covers a five year period. The agenda of the Park Authority for hill farming is on paper immaculate and the aspirations are all eminently worthy of support. For instance, a report of 2006 on 'Measuring the state of farming and land management' produced for the 'State of the Park' annual reporting process, states that "the vision for conserving and enhancing the National Park cannot be done without farmers and land-managers." Although the concern of the report is to identify any changes quickly and, "to take appropriate action to support sustainable farming activity in the Lake District," there is no indication of what the action will be, rather it is a call to collect information about changes in agri-environment coverage, changes in the farming economy and changes in the structure of farming. There are, however, no blueprints for change or the significant development of new initiatives to tackle the known and well-rehearsed problems, not least because the resources are not perceived to be readily available – even though there is rural development and other funding available to assist. The track record would suggest that the National Park Authority feels it is able to do very little about the needs of the hill farming sector as much as it appreciates the importance of the issues.

Certainly direct intervention through owning farms has not been a substantial part of the National Park's agenda. Until recently the Authority has owned just one hill farm, Beckside, at Sandwick, which according to some was probably in the 'wrong place' if the intention was for it to act as a showcase for sustainable farming. This notwithstanding, the Ivinson family who were its tenants for a number of years sought to do all the right things: they participated fully in environmental schemes; they kept Herdwick as well as Swaledale sheep, and developed a quality herd of Beef Shorthorn cattle. They also added value to their products by developing a business marketing their meats. They sought to take seriously the intention that the Park originally had of educating the public about traditional hill farming, but perfectly reasonably identified the need to make it pay by developing a tearoom which would have also provided environmental interpretation, permission for which was refused. The refusal was made on the grounds that it would be wrong to generate any additional traffic into an isolated area – even though there are already an estimated 100,000 people passing through the area every year.

In March 2007 it emerged that the Authority would put Beckside on the market in the summer of 2007 with a probable price tag of £1.5m.[9] Bob Cartwright, director of park services, justified the decision to sell the farm on the grounds that it did "not contribute to national park purposes." It surely could have been used in such a way as to fulfil the National Park purposes of encouraging public enjoyment and understanding of a special landscape and it might even have been used to demonstrate to the farming community aspects of the new agenda for agriculture.

So, if the National Park is retreating from its very minor role as an owner of farms, what might it do for traditional fell farming and Herdwick sheep keeping? True, the Park Authority's area rangers can give some support to farmers trying to avoid (or sorting out) problems that arise as a result of visitor pressure; the estates teams often provide practical help and their officer who deals with farming and conservation issues can keep a watching brief and maintain some input about developments in the hill farming sector. Unlike the National Park Authority in the Yorkshire Dales, however, the Lake District did not opt to become the vehicle for the delivery of the area's ESA scheme. This could have made a huge difference and seen a new orientation for the Lake District National Park – one that recognised the importance of its most important land managers.

An example of what might be done comes from one of the newest UK National Parks, that of Loch Lomond and the Trossachs, which has pioneered the use of a Land Futures Action Plan. Land Futures programmes are about participatory planning for land managers. In one example, that for Glen Dochart and Strathfillan, the process involved the following: an area of over 33,000 hectares was identified; the area was described and the area's land managers were identified (22 in all – made up of private, public and community organisations); their land management activities were described as were their diversifications and their management of bio-diversity and of Natural Heritage Sites; the impact of climate change was looked at and public access, tourism, culture and heritage issues were briefly described. An analysis of the strengths, weaknesses, opportunities and threats in the area was carried out by the land managers and then a vision statement was produced which assembled the main hopes for the future.

This vision covered the following themes and priorities: 1) sustainable land-based industries, 2) access and visitor management of access, 3) renewable energy and eco-system services, 4) rural skills and education and 5) information, communication and networking. For each theme actions which might progress the priorities were identified and the organisations responsible for taking the actions were listed, including the land managers themselves. Within each theme actions which should be taken forward over a three to nine month period were identified by the land managers. In total 35 actions were identified across the range of themes with, for example, the priority action in sustainable land-based businesses theme being to investigate the potential for local food and local forest product initiatives within the local tourist industry.

It is pleasing to report, however, that more recent policy making for the Lake District seems to recognise the need for action. In the National Park's Local Development Framework Core Strategy produced in May 2008 there is an array of suggestions. These include working with farmers to promote their conservation role; encouragement of local shops and supermarkets to sell local products; the promotion of farmers markets and local abattoirs; the promotion of farming as a career for young people and even the promotion of the re-use of farm buildings for income generating uses. Thankfully the Park Authority now makes it clear, "It is essential that we support and promote sustainable farming activity in the National Park. Together with out partners, we must find common ground within the needs of a vibrant farming community and the benefits to society of a sustainably managed landscape." To adopt the Land Futures approach to implement these suggestions using Rural Development Programme for England funding would make complete sense.

This is surely an approach that should be replicated throughout the hill farming areas of the UK. Nowhere would its application be more timely than in the Lake District. It may indeed be the case that techniques of this sort will be relevant as the Lake District National Park Authority and a wider partnership now develops a bid to inscribe the Lake District as a World Heritage Site. This will involve the development of a new approach to sustaining hill farming in the area which really heralds the possibility of the development of a new approach to the management of the National Park. Much will depend on the energy and commitment demonstrated by the organisations behind the bid and whether the National Park Authority can change its culture from theory to practice. Properly handled and seriously implemented it could represent a new departure in the management of the National Park and its cultural landscape. It will be important, however, that people understand the nature of that case.

Cultural Landscape: Towards a World Heritage Site?

A partnership of Cumbrian and regional organisations is working towards making a case for the Lake District to be a candidate for inscription by UNESCO as a World Heritage Site under a 'cultural landscape' classification. A cultural landscape is defined as a landscape which reflects the 'combined works of nature and man' and which can be seen also as a place that in an exemplary way illustrates the 'evolution of human society and settlement over time.' The fundamental, underpinning issue concerned in all this is 'time depth' – i.e. the area's long-standing history of continuous upland pastoralism going back for at least 900 years.

A consultant carrying out an early scoping study of the Lake District's case identifies the accessibility and openness of the Lake District upland landscape as one of the key issues and writes of the importance of:

"The continuing use of uplands by hill farmers for communal sheep grazing [which] has perpetuated a largely unenclosed terrain across which there is access both *de facto* and *de jure* and a web of rights of way…"

Further, to use the language of UNESCO which determines these matters, the Lake District shows outstanding "time depth and continuity", not least round the related point of the "longevity of the tradition of managing land" and a "farming regime based on year-round grazing on uplands managed in common with indigenous breeds among which the Herdwick is outstanding."[10]

The forces that link the work of nature and of man are the key things in a 'cultural landscapes' approach. Key issues in cultural landscape status are the tests of authenticity and integrity. As Susan Denyer puts it, "what is now coming strongly into focus is the impact of people on the environment, and the impact of the environment on people's livelihoods." In short, a major concern is about sustaining living landscapes. Denyer points out that all landscapes are in some senses cultural landscapes, but only a few have the required 'time depth' to qualify for World Heritage Site status. She points out that globally there are a very few landscapes which have a continuity of cultural practices stretching back for a thousand years. She enumerates some of these as follows: south-eastern Turkey, northern Yunnan in western China, the High Alps in Switzerland, the northern Caucasus in Georgia. The Lake District sits alongside these. They are all areas where a distinctive way of managing the land has persisted over a long time span. But there are vulnerabilities about these areas – the Lake District amongst them – which should be recognised and about which action should be taken. Watercourses, wall systems, communal management of extensive areas of common fell grazing – all these and more are things that are likely to be neglected if the economic viability of farms is low. Denyer concludes that:

"The overall vulnerabilities of the landscape as a whole are all too apparent. The number of farmers in the Lake District has dramatically fallen over the last four hundred years... This decline in numbers of people living and working on the land means that there is now an inadequate structure to allow communal management of some of the key elements of the landscape, such as walls, hedges, watercourses and grazed common fells. In many valleys the bones of the landscape pattern survive but they are hanging on by a thread, as the social and economic systems that supported them are weakening.

"Adding to the overall vulnerabilities of key cultural and social elements is the very recent sharp decline in the viability of farming, and the way grant-aid has been targeted towards the natural elements in the landscape rather than the cultural elements – because the identification of natural elements is many years ahead of work on the identification of cultural elements."[11]

What are the cultural elements of the Lakeland landscape? The cultural landscape case is not (or at least should not be), however, remotely about Peter Rabbit and friends and their creator; nor has it got much to do with the fact that the great Victorian sage John Ruskin lived for many years overlooking Coniston Water. It may have a lot more to do with what Wordsworth saw and appreciated so strongly in the working landscape, full of the work of people as well as of the forces of nature as is so amply demonstrated in Book Eight of *The Prelude* which Wordsworth entitled "Retrospect – Love of Nature Leading to Love of Mankind." Wordsworth saw and marvelled at the work of farmers and of shepherds and saw the connection through pastoralism between the work of people and the work of nature, as this extract from his poem so clearly reveals:

> *A Shepherd in the bottom of a Vale*
> *Towards the centre standing, who with voice,*
> *And hand waved to and fro as need required*
> *Gave signal to his Dog, thus teaching him*
> *To chace along the mazes of steep crags*
> *The flock he could not see: and so the Brute*
> *Dear Creature! With a man's intelligence*
> *Advancing, or retreating on his steps,*
> *Through every pervious strait, to right or left,*
> *Thridded a way unbaffled; while the Flock*
> *Fled upwards from the terror of his Bark*
> *Through rocks and seams of turf with liquid gold*
> *Irradiate, that deep farewell light by which*
> *The setting sun proclaims the love he bears*
> *To mountain regions.*

So powerful is Wordsworth's contribution that Professor Peter Fowler even suggests that the Lake District, as well as being "an organically evolved landscape" of a continuing kind, qualifies under the World Heritage Site criteria as an "associative cultural landscape." He says this because the Lake District was in fact the place where the "revolutionary concept of the aesthetic of landscape appreciation was conceived and developed." Fowler argues that "western-educated eyes cannot look at landscape at all except through Lake-District-tinted spectacles."[12] Even if this overstates the case somewhat, it is definitely the case that the 'national property' argument that Wordsworth espoused in the early nineteenth century and the inspiration of his observation about the interactions of people and nature inspired much of the subsequent thinking about the management of areas of high landscape value.

An important part of the case being put together for the Lake District would revolve round the production of a convincing plan to manage and interpret the cultural landscape for which the status is being sought. This management would, inevitably, have to concern itself with the maintenance of the fell farming system. The World Heritage Site management plan would need to establish a baseline by benchmarking the system including the contribution of Herdwick sheep. The state and disposition of fell farming would have to be established on a whole valley basis, valley by valley.[13] It would be necessary, therefore, for information to be shared about the National Trust farms with some assessment of their future management.

A similar piece of work would need to be done for the fell farm estate of United Utilities (which is significant in the Haweswater and Thirlmere areas) as well as of the smaller public holdings of Cumbria County Council and the Forestry Commission. The holdings of larger private landlords would need to be considered and attention would also have to be given to the likely fate of the numerous owner-occupied genuine fell farms such as Woodhow, Wasdale, which is currently on the market for £1.2 million after a short period under new ownership. All this would contribute to the baseline statistics against which future change would be measured.

The continuing management of common land would need to be a major consideration, both in relation to overall decision making through Commoners' Associations and the Commoners' Federation. There would need to be some assessment of the numbers of people and their level of skill in gathering the fells. There would have to be an audit of the state of traditional and native animal breeds involving an assessment of numbers and distinct bloodlines on the lines of the work being done by the Sheep Trust on 'heritage breeds' (which include the Herdwick and the Rough Fell) in 2007. An exercise should also be done as to the costs of using the European Rural Development Regulation's provision for making payments to breeds "in danger of extinction." This would be a way of keeping a direct and efficient instrument of appropriate support which would contribute to sustainable development in front of Defra in spite of their dogmatic resistance to payments based on headage.

Crucially there would need to be work done on succession issues, trying to establish where the future land managers will come from. Since this is very dependant on the general prospects of the sector, there would have to be the outlines of a system for identifying suitable young people in spite of prospects being poor. This system would ensure that interested young people were provided with training and mentoring opportunities in order that skills and traditional knowledge were transmitted. Attention would have to be paid to the housing opportunities for these young people and with imagination and commitment it might be possible to design a 'fell farming ladder' whereby young people could build their CVs and their savings in order to move progressively towards a rented farm of their own. This might often involve a part-time farm such as the threatened small National Trust farm of Hoathwaite at Torver combined with work for other people as a step on the way.[14]

It would also be helpful to develop a communication strategy to interpret fell farming to people more widely. Communication and 're-connection' programmes such as the work of the National Trust's Food and Farming officer, the Flora of the Fells programme developed by Friends of the Lake District, and the awareness-raising events for the public organised on farms by the Cumbria Farmer Network all need to become mainstream with resources being gathered from rural development sources to continue them. This work might sit alongside anything that the National Park Authority decided to do on the communication front, perhaps to replace the guided walks programme that lost its budget a couple of years ago. It would also be valuable to take forward work done by the Fells and Dales LEADER + Programme on the establishment of a Hill Farming Charter through a process of discovering with the farmers at the level of the individual farms, the key features of their delivery of both tradable products and public goods.

There would also be potential to develop one or more 'eco-museums' of the area's pastoralism.[15] This would take the form of a community-led initiative making accessible, in situ (rather than ex situ, in a special museum), the operation of the fell farming system through interpretation of sites and features (the farmhouses, the barns and byres, the hogg houses and the sheepfolds, the walls and wall systems, the intakes, outgangs and drove roads, the bridges, the open fells) and by opportunities for observation of the husbandry practices involved with keeping grazing animals, especially sheep on the fells. This might usefully include the provision of interpretative material (along with the availability of food and drink) in specially adapted farm buildings.

Guided walks (led by fell farmers) as well as self-guided trails could be developed and there could be provision of new events and celebrations alongside existing events such as shepherds' meets and shows. A programme of this sort has clear potential appeal for the growing number of people who are interested in cultural landscapes and eco-tourism. It would also provide the fell farming community with a sense that they were being properly valued and they might, within this initiative, find new markets for their fell farm products.

In the development of the case for the Lake District to become a World Heritage Site one can begin to see the beginning of a realisation that the keeping of Herdwick sheep on the fells is a matter of some significance. But incomes on fell farms remain low; Herdwick sheep numbers continue to decline; there are clear problems of succession on fell farms and concerns over the transmission of the necessary skills, knowledge and culture to the next generation. One hopes that World Heritage Site status will provide the fell farming sector in the Lake District with opportunities for additional financial support for the production of the significant public goods it delivers and also that it will provide greater opportunities for the creation of added value markets for fell farm products. The fortunes over the next few years of the Herdwick breed of sheep and of traditional Lake District fell farming generally will be a proxy for the health of the cultural landscape created round its keeping – and this will be the case irrespective of World Heritage Site cultural landscape status.

The growing interest at European level, for instance, in the application of the High Nature Value (HNV) farming concept could also make a great contribution to places like the Lake District with their "traditional agricultural landscapes." The HNV farming concept, according to a recent report had its origins, "in a growing recognition that throughout Europe, many of the habitats and species upon which we place high nature conservation value, and which are declining, have been created, and need to be maintained by, farmers and their farming practises."[16] It is further suggested that Member States could identify "Traditional Agricultural Landscapes" (TAL). Traditional fell farming in the Lake District – and the related keeping of Herdwick and other fell sheep – would surely qualify in the UK context, based on the criteria below, as a Traditional Agricultural Landscape:

"1) The existence of high aesthetic and cultural values; 2) the pursuit of a broadly traditional or locally adapted approach to management; 3) the presence of features, whose distribution is regionally and/or locally specific, which contribute to its aesthetic qualities and to its ecological integrity."[17]

Herdwick sheep-keeping is an ancient yet continuing phenomenon that has shaped the landscape fundamentally and has had significance beyond its local impact. As this account has shown it has led to the creation of a rich local culture of fell sheep farming and has led to crucially important considerations about the relationship between people and nature and the aesthetics of the countryside. It is also the case that it was in the Lake District, amongst other places, that the European approach to protected landscapes originated and has been developed – notably in National Parks operating in populated and farmed environments.

Traditional fell farming has also been right in the centre of the development of the popular desire for access to the countryside in what was the world's first industrial nation. As has been shown, public access to the countryside above the fell wall was highly congruent with Herdwick sheep keeping. The sheep keeping economy and its culture are central to the creation of the forces that led to the creation of the national park movement, with the anti-afforestation campaign of the inter-war years being centrally tied up with the Herdwick story. The story of the acquisition of fell farms by various benefactors and organisations like Lake District Farm Estates and the National Trust in the Lake District shows very clearly the role of traditional farming in the case for National Parks on the British model. Traditional farming has also made a major contribution to the livelihoods of Lake District communities. There is, however, a fragility in the fell farming economy that is readily apparent. It is clear that Susan Denyer has it right when she says that possible World Heritage Site designation offers:

"an ideal opportunity to consider the way forward and to identify what the key significances and vulnerabilities of the landscape are, what conservation options exist, what should be managed, how a holistic approach to management could be achieved and how residents, visitors and the multiplicity of agencies could collaborate to achieve these ends."[18]

The time has now come to say that the Lake District is too important to be left to market forces. Strong interventions by a wide range of stakeholders are required to protect the cultural landscape of the area. The Lake District remains and needs to remain as a working landscape in spite of all the forces that work against that. The time has come to say that there should be no more erosion of the fell farm base that delivers so much of the management of the landscape. Ways should be explored to ensure the strengthening of the labour force that carries out the necessary management of the land through the rearing of grazing animals. Every significant dale and massif in the Lake District should draw up a whole valley or area plan. The farmers should be facilitated to do this and their ideas for ways forward for the area should be brought to the centre of the debate. A proper management plan with an agreed agenda of actions should be drawn up. Resources should then be found for implementing the actions and this would have to be done by all the stakeholders working together.

There is little doubt that this is urgent work for there are big challenges on the horizon: ranging from Blue Tongue virus and the impact of climate change through very high fuel and animal feed prices to changes in agri-environment support and further change to the CAP. Maintaining the Herdwick flocks and the fell farming system needs to be an absolute priority for the Lake District. The area is not just 'a national property' it also has a resonance as a special place in world terms. The Herdwick sheep flocks that graze the fells and the people who manage them underpin the English Lake District's integrity and authenticity as a cultural landscape and do much to create the appeal of the area for the millions of people, world-wide, who cherish it.

Annex

Cumbrian Fell-going Flocks

Farms on which fell-going flocks of Herdwick are thought by the author to have been kept between 2001 and 2008. Farmers' names are given in brackets. It should be remembered that a number of these farms have several stocks of Herdwick sheep going on different fells and heafs. I apologise to anybody who has been missed off the list.

Stone Ends, Mungrsidale (Clark)
Middle Row, Threlkeld (Tyson)
Millbeck Hall, Bassenthwaite (Brownrigg)
Ashness, Borrowdale (Cornthwaite)
Fold Head, Watendlath (Richardson)
Hollows, Borrowdale (Fearon)
High Lodore, Borrowdale (Weir)
Nook, Borrowdale (Jackson)
Yew Tree, Borrowdale (Relph)
Stonethwaite, Borrowdale (Gill)
Thorneythwaite, Borrowdale (Bland)
Chapel House, Borrowdale (Weir)
Seathwaite, Borrowdale (Edmondson)
Seatoller, Borrowdale (Simpson)
Little Town, Newlands (Relph)
Low House, Newlands (Folder)
Skelgill, Newlands (Grave)
Gatesgarth, Buttermere (Richardson)
Syke, Buttermere (Kyle)
Wilkinsyke, Buttermere (Beard)
Cragg House, Buttermere (Temple)
Croft House, Buttermere (Stagg)
Rannerdale, Buttermere (Beard)
Miller Place, Lorton (Blair)
Swinside End, Lorton (Nicholson)
Thackthwaite, Loweswater (Cartmel)
Kirkhouse, Setmurthy (Norman)
Hatteringill, Mosser (Clark)
Mossergate, Mosser (Clark)
Howe Ghyll, Lamplugh (Richardson)
Smaithwaite, Lamplugh (Benn)
Mireside, Ennerdale (Vickers)

How Hall, Ennerdale (Hardisty)
Moor End, Ennerdale (Hinde)
How Side, Ennerdale (Maxwell)
Hollins, Ennerdale (Rawling)
Strudda Bank, Beckermet (Holliday)
Thornholme, Calder Bridge (Crayston)
Wind Hall, Gosforth (Bolton)
Whinnerah, Gosforth (Nicholl)
Scalderskew, Gosforth (Ellwood)
Thistelton, Gosforth (Hindmoor)
Church Style, Nether Wasdale (Knight)
Easthwaite, Nether Wasdale (Steele)
Bridge End, Santon Bridge (Phizacklea)
Woodhow, Wasdale (Scrivenor)
Ghyll, Nether Wasdale (Shuttleworth)
Bowderdale, Wasdale Head (Naylor)
Wasdale Head Hall, Wasdale Head (Hodgson)
Middle Row, Wasdale Head (Naylor-Lopez,)
Burnthwaite, Wasdale Head (Race)
Howbank, Birkby (Matterson)
Crag Farm, Birkby (Matterson)
Charles Ground, Corney (Jenkinson)
Welcome Nook, Corney (Ellwood)
Low Place, Mitredale (Steele)
Gill Bank, Eskdale (Temple)
Spout House, Boot (Postlethwaite)
Wha House, Eskdale (Crowe)
Penny Hill, Eskdale (Jackson)
Taw House, Eskdale, (Fox)
Butterilket, Eskdale (Harrison)
Dalegarth, Eskdale (Stanley)

Field Head, Eskdale (Benn)
Howes, Eskdale (Baines)
Black Hall, Ulpha (Temple)
Cockley Beck, Seathwaite (Wrathall)
Troutal, Seathwaite (Bradley)
Tongue House, Seathwaite
    (Clegg and Bennett)
Turner Hall, Seathwaite (Hartley)
Wallabarrow, Ulpha (Chinn)
Old Hall, Ulpha (Hoggarth)
Kiln Bank, Ulpha (Hoggarth)
Hazel Head, Ulpha (Angus)
Baskell, Ulpha (Ellwood)
Pike Side, Ulpha (Askew)
Crosbythwaite, Ulpha (Harrison)
Woodend, Ulpha (Crowe)
Hoses, Dunnerdale (Gorst)
Sella, Dunnerdale
    (Kirkby and Longworth)
Greenbank, Broughton Mills (Hannah)
Fenwick, Thwaites (Johnson)
Thwaite Yeat, Thwaites (Troughton)
Boadhole, Thwaites (King)
Troughton Hall, Woodland (Cooper)
Heathwaite, Torver (Lancaster)
Hoathwaite, Torver (Wilson)
Tilberthwaite, Coniston (Wilkinson)
Boon Crag, Coniston (McCartney)

Coniston Hall, Coniston (Wilson)
Fell Foot, Little Langdale (Benson)
Birk How, Little Langdale (Birkett)
Middle Fell, Great Langdale (Toms)
Stool End, Great Langdale (Rowand)
Wall End, Great Langdale (Edmondson)
Bayes Brown, Great Langdale (Rowand)
Millbeck, Great Langdale (Taylforth)
Robinson Place, Great Langdale
    (Edmondson)
Harry Place, Great Langdale (Benson)
Troutbeck Park, Troutbeck (Tyson)
Brockstones, Kentmere (Dickinson)
West Head, Thirlmere (Bland)
Town Head, Grasmere (Hodgson)
Brimmer Head, Grasmere (Powell)
Stybeck, Thirlmere (Hodgson)
Thirlspot, (Gaskell)
Knott Houses, Grasmere (Bland)
Ann's Cottage, Dockray (Wilson)
Penfold, Dockray (Nicholson)
Ulcat Row, Matterdale (Potter)
Dowthwaite Head, Matterdale (Weir)
Noran Bank, Patterdale (Barker)
Deepdale Hall, Patterdale (Brown)
Glencoyne, Glenridding (Hodgson)
Gillside, Glenridding (Lightfoot)
Howe Green, Hartsop (Wear)

Smaller Cumbrian Flocks

Farms within Cumbria with small and/or non-fell going stocks at many of which Herdwick tups have been registered in recent years. This list includes some tup breeders of repute.

Hollins, Dockray (Bennett)
Crookabeck, Glenridding (Bell)
Little Cross, Arkleby (Hill)
Angerton, Sandale (McFarlane)
Torver (Barr)
Wythop Hall, Wythop (Emmott)
Briar Bank, Cockermouth (Folder)
Moor, Keswick (Tyson)
Barfield, Bootle (Hartley)
Skelsceugh, Frizington (Tyson)
Climb Style, Woodland (Thompson)
Knott End, Grizebeck (Campbell)
Ellermire, Grizebeck (McKinney)
The Galleon, Hawkshead (Harryman)
Drigg Cross, Gosforth (Grave)
Hawthorn Cottage, Holmrook (Watson)
Greenmount, Crooklands (Grisedale)
Fornside, St Johns in the Vale (Hall)
Stoddah, Matterdale (Clark)
Winscales West, Egremont (Hogg)
Home Farm, Thwaites (Braithwaite)
Lorton Park, Lorton (Nicholson)
Brownrigg, Matterdale (Rebanks)
Laithes, Greystoke (Beaty)
Farlam, Brampton (Elliot)
Crag Hall, Thwaites (Graveson)
High Dyke, Mosser (Heslop)

and three from North Yorkshire:
Middlesmoor, North Yorkshire (Bowles)
Manor House, Ingleton (Metcalfe)
Braime House, North Yorkshire (Hinde)

Notes and References

Chapter 1
Origins and Development of the Herdwick Breed

1 William Wordsworth, *Guide to the Lakes*, (fifth edition 1835, 1977 imp.) p.26
2 I. D. Whyte, 'The Dark Age Landscape' in *The Lake District: Landscape Heritage*, ed. W. Rollinson (London 1989), p.65 and N. Higham, 'The Scandinavians in North Cumbria' in *The Scandinavians in Cumbria*, eds. J. R. Baldwin and I. D. Whyte, (Edinburgh 1985) p. 43
3 C. M. L. Bouch and G. P. Jones, *The Lake Counties, 1500-1830: a social and economic history* (Manchester 1968 imp.) appendix one, 'Herdwick Sheep', p. 348. In Norwegian 'sau' means sheep. It is interesting to note that the Spaelsau sheep of Norway are often called 'villsau' or wild sheep and referred to as being 'utegangarsau' – echoes of which can clearly be found in the dialect word 'outgang' (way out to the fell) and in the notion of 'fell-going sheep' that was historically a synonym for Herdwicks.
4 I. D. Whyte, 'Shielings and the Upland Pastoral Economy of the Lake District in Mediaeval and Early Modern Times' in *The Scandinavians in Cumbria*, eds. J. R. Baldwin and I. D. Whyte, (Edinburgh 1985) p.103
5 David Kinsman, 'Herdwick origin is mystery' in *Westmorland Gazette*, 8 October 2004. See also his *Black Sheep of Westmorland* (Windermere 2001). He notes on p.21 that the Herdwick constitutes a sheep type all of its own and comments that because it is long tailed, "any association with the Norse is most unlikely."
6 M. L. Ryder, *Sheep and Man* (London 1983), pp.460-2. It is certainly still the case that the Herdwick (without any tail shortening) has a shorter tail than the Swaledale. Long tails in Herdwick sheep are frowned upon – undue length often being ascribed by shepherds to a genetic contribution from a black-faced sheep.
7 M. L. Ryder, *Sheep and Man*, p.769. Perhaps the best summary has been provided by W. Pearsall and W. Pennington, *The Lake District* (London 1978 edn.) pp.272-3 where they concede the likelihood of a black-faced influence on a "small white-faced hairy sheep with a high haemoglobin A frequency."
8 One breed, the Silverdale (sometimes called the Limestone or Farleton Knott) became extinct around the First World War. This is well-documented in F. W. Garnett's *Westmorland Agriculture, 1800-1900* (Kendal 1912).
9 Bailey and Culley's book on Cumberland can be found in the reprint of the same authors' *General View of the Agriculture of Northumberland*, ed. D. J. Rowe, (Newcastle upon Tyne, 1972). The quotations are from pp.245-247. This volume also contains A. Pringle's *General View of the Agriculture of the County of Westmorland*.
10 Whether all Herdwick tups looked like this at the time is, of course, debatable.

11 Clarke's observations are reprinted in Norman Nicholson's anthology, *The Lake District*, (Penguin 1978), p 321. Nicholson points out that the passage "is full of misunderstanding and inaccuracies," but it has become "part of the folklore of the Lakes."
12 Bailey and Culley, *General View of the Agriculture of Cumberland*, p.215
13 F. W. Garnett, *Westmorland Agriculture*, (Kendal 1912), p.18
14 Lord Ernle, *English Farming Past and Present*, (London 1927), 4th edition, p.181
15 William Youatt, *Sheep: their breeds, management and diseases* (London 1837), p.278
16 Bailey and Culley, *General View of Agriculture of Cumberland*, pp. 247-249
17 William Dickinson, *Essay on the Agriculture of West Cumberland*, (London and Whitehaven 1850), p.61
18 See for instance, *Whitehaven News*, 15 September 1870. On offer that year were 200 four year old wethers, 220 ewes and 70 lambs.
19 Blake Tyson (ed.), *The Estate and Household Accounts of Sir Daniel Fleming of Rydal Hall, 1688-1701*, (Cumberland and Westmorland Antiquarian and Archaeological Society, 2001) p.216
20 F. W. Garnett, *Westmorland Agriculture* p.162.
21 Quoted in Garnett, *Westmorland Agriculture*, p.156
22 William Dickinson, *The Farming of Cumberland* (1853) p.60. Dickinson also states that such a flock was kept successfully by Henry Hodgson on the fells above Ousby on the Pennines.
23 William Whelan, *History and Topography of the counties of Cumberland and Westmorland* (Pontefract 1860) p.67
24 'Herdwick Sheep' in *The Field*, 27 September 1873. A copy of this article was sent to me shortly before her death by the late Mrs. Susan Johnson of Ravenglass (the daughter of H. H. Symonds) along with other material she had collected on Herdwick sheep over the years. There is a picture of a prize winning Herdwick tup, the 'Old General', from the 1870s that is of a similar stamp, in the possession of Wm. Wilson.
25 H. D. Rawnsley, 'Joseph Hawell, a Skiddaw Shepherd' in *Lake Country Sketches*, (Glasgow 1903), p.152. Joseph's father was from Longlands, Uldale, and had been given ten gimmer lambs by his father.
26 H. D. Rawnsley 'A Crack About Herdwick Sheep' in *By Fell and Dale at the English Lakes* (Glasgow 1911) p.47
27 The Keswick May Fair – when tups hired over the winter are returned – still takes place on the Thursday after the third Wednesday in May. In the mid-nineteenth century the autumn tup fair at Keswick seems to have been held much closer to tup time in early November: see for instance the diary of the Rev. Basil R. Lawson (ed. Margaret Armstrong), *Thirlmere: across the bridges to chapel, 1849-1852* (Keswick 1989) where in 1850 "tup fair at Keswick" took place on 2 November. The backend of the year 'Keswick' Tup Fair now takes place on the first Saturday in October at Cockermouth Auction Mart.
28 Since the re-launch of the Fell Dales show on land near the 'King George IV' in Eskdale in 2003 (there was no show in 2001 due to foot and mouth disease and no show in 2002 due to the continuing complexities of 'bio-security'), there has been a local effort to collect information about the history of the show.

29 Rev. T. Ellwood, 'The Mountain Sheep: their origin and marking' in *Transactions of the Cumberland and Westmorland Antiquarian and Archaeological Society*, Old Series, Vol. XV, 1899. Ellwood was a very strong believer in the Norse origins of Herdwick sheep, but as has often been pointed out, the similarity between terminology used in Cumbria and in Scandinavia says more about the common origins of the shepherds rather than the sheep. The word 'mountain' is still used to describe the draft ewes at some sales, for instance, Broughton-in-Furness.

30 William Dickinson, *Essay on the Agriculture of West Cumberland*, (London and Whitehaven 1850) p.60. He commented that this practice of cross-breeding "meets more extensive patronage every year, and is a rather solitary but encouraging proof of progress in the annals of mountain sheep farming."

31 It was well established from the early nineteenth century that the older portion of Herdwick ewes would be drafted from the fells and put to Down and Longwool tups: for instance, the Leicester, the South Down and later the Wensleydale and Teeswater were all used. These crossing tups were known as 'Mugs.' Garnett mentions that in 1885 there were sheep classes at Westmorland County Show for Leicester, Longwool or Down (all 'mugs') alongside Fell, Herdwick and Horned Crag sheep.

32 F. W. Garnett, *Westmorland Agriculture*, 1912, pp.161-62. There were also classes for Herdwick sheep when the Royal Show, which in those days moved round the country, was in the north: e.g. Carlisle 1855, Newcastle 1864 and Liverpool 1877.

33 *The Annals of a Quiet Valley by a Country Parson*, edited by John Watson, (London 1894), p.180. It should be noted that many contemporary writers described Herdwick sheep as 'herdwicks': i.e. without a capital H.

34 A. Pringle, *General View of Agriculture of Westmorland*, p.327. The same words are used by John Housman in his *A Topographical Description of Cumberland, Westmorland, Lancashire and a part of the West Riding of Yorkshire* (Carlisle 1809), p.101. In a section on Westmorland Housman states that the breed of sheep on the mountains "is either native, or a cross with the Scotch rams."

35 A. Humphries, 'Agrarian Change in East Cumberland,1750-1900' (University of Lancaster M.Phil thesis 1993) p.490

36 There is still a great deal of informal visiting to see other breeders' tups in the weeks before the backend tup sales.

37 General information on the Nelson family from Irvine Hunt, *Lakeland Yesterday*, volume 1, (Otley 2002) pp.82-3. See also Frank Carruthers, 'The Man who loved Herdwicks' in *Lore of the Lake Country* (London 1975), pp.120-129

38 Herdwick Sheep Breeders' Association, *Flock Book of Herdwick Sheep*, Vol. 1, (Penrith 1920) p.31

39 See entry on Gatesgarth in HSBA, *Flock Book of Herdwick Sheep* (1920); also conversation with Syd Hardisty of Ennerdale, whose father was shepherd at Gatesgarth with Ned Nelson Jnr. for eighteen years. See also R. H. Lamb's obituary of him in *Cumberland and Westmorland Herald*, 15 September 1934

40 Quoted in Andrew Humphries, 'Agrarian Change in East Cumberland, 1750-1900,' (University of Lancaster M.Phil. thesis 1993), p.486

41 Gate's *New Shepherd's Guide*, (Cockermouth and Lancaster 1879), p.357. There were Herdwick sheep at Naddle until the late 1980s. The current tenant Eddie Eastham told me in 2008 that ideally Herdwicks would replace the Swaledales on the farm's higher fells, with an increased number of Cheviots on the lower ground. The quotations from the 'Essays on Herdwicks' are from pp.490-3
42 See Ian Whyte, *Transforming Fell and Valley: landscape and parliamentary enclosure in north west England* (Lancaster 2003), pp.33-35. Whyte's book is mainly based on Westmorland and Lancashire sources.
43 J. H. Clapham, *An Economic History of Modern Britain: the early railway age, 1820-1850* (Cambridge 1926) p.15
44 S. D. Stanley-Dodgson, 'The Herdwick Sheep. Their Origin and Characteristics' in Herdwick Sheep Breeders' Association, *Flock Book of Herdwick Sheep*, Volume 1, (Penrith 1920); Crayston Webster, 'The Farming of Westmorland' in *Journal of Royal Agricultural Society of England*, 1868, quoted in A. Humphries, 'Agrarian Change in East Cumberland,' p.487
45 Garnett, *Westmorland Agriculture,* p.158
46 R. H. Lamb, *Herdwicks: Past and Present*, (Penrith 1936) p.7
47 It should be noted that when a Herdwick ewe is crossed with a ram of any other breed, the non-Herdwick fleece colour is always dominant – the only exception is when the other breed is a Swaledale or other sort deriving from the Black-faced Heath Sheep when the outcome will share characteristics of both breeds, i.e. it will have spots, small horns and often a grey rather than a black or white face.
48 See Graham Murphy, *Founders of the National Trust* (National Trust 2002 edn.), chapter on Hardwicke Rawnsley.
49 Papers given to me by the late William Bowes of Fenwick, Thwaites; these papers include the invitation to the first meeting sent to Thomas Bowes his father; a printed report of the 6 September meeting from the *Keswick Guardian* and a handbill calling people to the first General Meeting which was to be held on 1 December 1899 at the Parish Room, Keswick.
50 It is interesting to note that the HSBA was one of the first breed associations in the hill sheep sector and pre-dated the Swaledale and Rough Fell associations by several years.
51 H. D. Rawnsley, 'A Crack about Herdwick Sheep' in *By Fell and Dale at the English Lakes*, (Glasgow 1911), pp.68-72
52 See C. Bryner Jones, ed. *Livestock of the Farm*, Vol IV, Sheep, (London n.d, ?1912), p.7. Garnett in *Westmorland Agriculture*, p.174 gives an estimate that in 1862 there were about 2 million pounds of Herdwick wool, at 3.5lbs per fleece this means that there were in excess of half a million Herdwicks.
53 Garnett, *Westmorland Agriculture*, p.162. These semi-polled ('cowed') tups are still common and very acceptable.
54 There is little doubt that Herdwick Billy earned his soubriquet – he is widely considered to have been the driving force behind the early years of the Association. I have often been told of his outstanding commitment to the cause of Herdwick sheep.
55 The ewe figures given are those in the *Flock Book*. I think they must be regarded as minimum numbers in many cases and may just be the landlord's flock number in some cases and they will exclude shearlings. Flocks in this era would also contain numerous wethers.

56 Until recently there was a tradition of showing and having available for hire pens of five tups – which, given the normal complement of 50 ewes to each tup, would meet the needs of flocks of this common size.

57 It was also recorded by R. H. Lamb in the *West Cumberland News*, 3 April 1942 that Thomas Bowes also took over part of the Broadgate stock of Herdwicks owned by Sir William Lewthwaite which the Lewthwaites had held since 1657.

58 HSBA, *Flock Book of Herdwick Sheep*, 1921. The Harrisons were at Butterilket in 1920 also, as they still are today, but they did not join the Association until 1929. The Grave family claims to have been at Skelgill, Newlands, since 1347 even though the first written record of that is 1647: see Molly Lefebure, *Cumberland Heritage* (London 1974 edn.) p.99. The Rawling family is mentioned in the Lamplugh parish records going back to the sixteenth century.

59 HSBA, *Flock Book of Herdwick Sheep*, 1921.

60 *Penrith Herald*, 27 March 1920

61 Conversation with Pat Temple, Ulpha, 2005. The Temples were at Standing Stones, Kinniside, for the years 1936-1940. I have also been told that it was quite common just to view the land and the agricultural buildings on a potential farm and not even to bother to look in the house.

62 Information from William Wilson, Jnr. 2007

63 R. H. Lamb, *Herdwicks Past and Present*, p.12

64 Jerry Richardson recorded in his diary entry for 27 February 1929 that the "last of the old Seathwaite ewes died."

65 Keswick May Tup Fair Minute Book, 13 May 1922. It is interesting to note that what became the Swaledale sheep breeding community was also going through similar debates at this time, though in their case the issue was exactly how wide the type should be. A meeting at Barnard Castle proposed to set up a Blackfaced Dales-bred Flock Book. A rival meeting at Kirkby Stephen attended by breeders from Swaledale and Wensleydale, as well as Westmorland, favoured the use of the Swaledale name, though the majority were also happy with the Dales-bred Swaledale or the Swaledale Dales-bred. Mr. J. Bailey of Musgrave, however, dissented and said that "Dales-bred means roughs, Scotch or anything. Call them Swaledales! (Laughter)." See *Mid-Cumberland and North Westmorland Herald*, 28 February 1921. A separate Dalesbred Sheep Breeders' Association was formed in 1926, as was the Rough Fell Association.

66 Mrs. Heelis was amongst the number at this meeting and the subsequent one.

67 R. H. Lamb, *Herdwicks: Past and Present,* pp. 28-29

68 E. M. Ward, *Days in Lakeland*, (London 1948 edn.), p.77

69 *Beatrix Potter's Letters*, ed. J. Taylor (London, 1989) p. 367

70 Jonathan Brown, *Agriculture in England: a survey of farming, 1870-1947* (Manchester 1987), chapter 7, 'The Second World War.'

71 HSBA Minute Book, Council Meeting, 20 May 1943. Mrs. H. B. Heelis was in the chair for this meeting.

72 Sir Clement Jones, *A Tour in Westmorland* (Kendal 1948), p.19. There had also been bad winters in March 1933, January 1936 and January 1940.

73 Hartsop Hall is often thought to be the first major Lake District fell farm to switch to the Swaledale – which it did around the First World War.

Notes and References

74 E. M. Ward, *Days in Lakeland*, (London 1948 edn.) pp.79-80
75 Keswick May Tup Fair Minute Book, entry for 30 March 1935. Book in possession of W. D. Tyson, Middle Row, Threlkeld.
76 R. H. Lamb, *Herdwicks: Past and Present*, p.11.
77 Handbill of the sale in possession of the Rawling family, Ennerdale. Some of these breeding strategies were obviously carried out deliberately and as 'one-offs', given that it was known for some time that this change of holdings was going to happen. But even so it illustrates that the use of the Swaledale increased returns from sheep sales.
78 J. F. H. Thomas, *Sheep*, (London, 1945) pp.30-31.

Chapter 2
Fells and Heafs

1 According to the Royal Commission on Common Land, 1955-1958, Westmorland had 130,066 aces of common and Cumberland had 110,357 acres: see W. G. Hoskins and L. Dudley Stamp, *The Common Lands of England and Wales* (London 1964 imp.) p.198. There is also a great deal of privately owned fell which is open to other fells, i.e. it has no effective boundaries which can turn sheep.
2 Quoted in Garnett, *Westmorland Agriculture* (1912), p.165
3 Garnett, *Westmorland Agriculture*, p.163
4 A. M. Armstrong and others, *The Place-Names of Cumberland*, Part II, (Cambridge 1950) p.343
5 *Oxford English Dictionary*, 2nd edition (2004 imp.)
6 Thomas Denton, *A Perambulation of Cumberland 1687-1688*, edited by A. J. L. Winchester in collaboration with M. Wane (Surtees Society 2003) p.79
7 Garnett, *Westmorland Agriculture*, p.163
8 Entries from *A Glossary of the Words and Phrases pertaining to the Dialect of Cumberland* by W. Dickinson, edited by E. W. Prevost (London and Carlisle 1879)
9 Susan Johnson, 'Two Duddon Farms, Thrang and Hazelhead' in *Cumberland and Westmorland Antiquarian and Archaeological Society* (New Series), Vol LXI, Kendal 1961, p.242 and pp.245-246
10 W. Gilpin to Lonsdale, 29 July 1696 in D. R. Hainsworth (ed.) *The Correspondence of Sir John Lowther of Whitehaven*, 1693-1698 (London 1983), letter 296.
11 Dickinson papers, West Cumbria Record Office
12 Blake Tyson (ed.), *The Estate and Household Accounts of Sir Daniel Fleming of Rydal Hall, Westmorland, 1688-1701* (Cumberland and Westmorland Antiquarian and Archaeological Society) pp.256 and 262. Tyson in note 206 asserts that Fleming was buying sheep heafed to Kentmere Hall land. It may be the case that Fleming let the farm in the following April: on 4 April 1699 he gave three shillings to sheep viewers at Kentmere.
13 Garnett, *Westmorland Agriculture*, pp.163-4.
14 T. W. Thompson, *Wordsworth's Hawkshead*, edited by Robert Woof (London 1970), pp.277-278

15 Quoted in Andrew Humphries, 'Agrarian Change in East Cumberland, 1750-1900', (University of Lancaster M.Phil. dissertation, 1993), p.498. See p.499 for details of sizes and compositions of other landlord's flocks on the Greystoke Estate at various dates in the nineteenth century.
16 H. H. Symonds, *Walking in the Lake District,* new edition revised by Susan Johnson, (London and Edinburgh 1962), appendix D, pp.320-1. This agreement is notable because it does not contain any wethers.
18 Article on Middlefell in *Cumbria Farmer*, February 2008. There is an account of 'Livering Day at Middlefell Farm' by M. Blake in *Cumbria: Lake District Life*, August 1976
18 Information from Nick and Tracey Gill, tenants at Stonethwaite, 2008.
19 *National Trust Properties*, March 1966, p.58
20 The three viewers currently for the National Trust are David Bland (Thirlmere), Andrew Nicholson (Lorton) and Anthony Hartley (Seathwaite).
21 The viewers at Tilberthwaite in 1986 were Ernest Wilson and Hugh Parker (for the National Trust) and Noel Hodgson and John Richardson (on behalf of the outgoing tenant).
22 Royal Commission on Agriculture, report on the County of Cumberland, C-7915-I, 1895, p.9
23 R. H. Lamb, *Herdwicks,* p.8
24 Wethers are also thought to eat coarser vegetation. English Nature carried out a grazing experiment using wethers in the central daleheads from the late 1990s. There is still an agreement to keep wethers on Burnthwaite's open fell at Wasdale Head, designed it seems to prevent incursion from other flocks. Keeping wether flocks largely died out in the 1960s when it began to make more sense to substitute ewes for wethers. The late Noel Hodgson of Bayes Brown, Great Langdale, told me that when Isaac Thompson retired from West Head in the late 1930s nine lorry loads of wethers were taken into the auction. When assessing what has happened to sheep numbers over the years, it is important to remember that ewes have displaced wethers.
25 Quoted in John L. Jones, 'And then there's the snow too' article on Millbeck, Great Langdale, in *Farmers' Weekly*, 2 March 1984
26 Cited in Charles Bowden, *The Last Shepherds* (London 2004) p.153
27 *The Diary of William Fisher of Barrow, 1811 to 1859*, ed. W. Rollinson and B. Harrison (University of Lancaster, 1986)
28 See Margaret E. Shepherd, *From Hellgill to Bridge End: aspects of economic and social change in the Upper Eden Valley,1840-1895* (Hatfield 2003) pp.164-165 and Bob Orrell, *Around and About Ennerdale* (Lamplugh 1997) p.82
29 M. L. Ryder, *Sheep and Man*, p.708 records that "Fly strike was rare on hill grazings in northern Britain before 1900." There is some evidence that susceptibility to blowfly strike may be a characteristic of the breed.
30 An elderly Great Langdale farmer is reported in 1983 as saying that sheep washing before clipping ceased before his time "because of t'maggots. Blowfly struck sheep before they were dry." See W. R. Mitchell, *Lakeland Dalesfolk* (Clapham 1983) p.43

31 W. T. Palmer, *The English Lakes*, (first published 1905, 1945 imp.) p.82. Blowfly maggots are known in dialect as either wicks or mawks and days when blowfly might strike as wicky or mawky. I have been given many examples of the struggle against the mawk in earlier times, for instance, by J. V. Gregg of Great Langdale and my former neighbour, the late John Roper of the Gill, Kinniside.

32 H. D. Rawnsley, *Months at the Lakes* (Glasgow 1906), p.106. About 1,300 sheep were clipped in a day.

33 Joseph Whiteside, *Shappe in Bygone Days* (Kendal 1904) pp.343-4

34 Graham Sutton, *Fell Days*, (London 1948) chapter 'Sheep and Sheep Enemies', pp.67-68 and see also H. D. Rawnsley 'A Crack about Herdwick Sheep' in *By Fell and Dale at the English Lakes*, (Glasgow 1911) pp. 55-57 for the seriousness of the maggot fly problem and how it dominated the fell shepherd's summer.

35 The story about Isaac Thompson is in W. R. Mitchell, *Lakeland Dalesfolk*, p.45. I have also heard this story from other sources. Other prices are from H. H. Symonds, *Afforestation in the Lake District* (London 1936) p.27

36 Gordon Ponsonby of Kinniside recalled to me some years ago that one job as a teenager that stuck in his mind around this time at the family farm of White Banks was that of collecting the wool in the spring from a group of sheep which had been smothered in a snow drift in Hole Gill on Lankrigg.

37 Andrew Humphries, *Agrarian Change in East Cumberland*, p.499

38 Brochure of sale (consulted in Whitehaven Library). Even the very small Wasdale church flock on High fell consisted of six wether shearlings and five wether hoggs alongside eighteen ewes, six gimmer shearlings and five gimmer hoggs.

39 W. T. Lawrence, 'Herdwick Sheep' in *Livestock of the Farm*, Vol. IV, Sheep, ed C. B. Jones (London n.d. ?1912), p.48

40 John Casson Sawrey, Copy Book, in possession of Duddon Valley Local Historical Association. I wish to thank Alan Linnitt for letting me see this. The clipping dates are by today's standards early, because there would otherwise have been problems with blowfly in an age before there were effective preventative treatments. In the first HSBA *Flock Book* he registered 25 tups in total, from the lambing times of 1914, 1915, 1916, 1917 and 1918.

41 A. J. L. Winchester, *Landscape and Society in Medieval Cumbria* (Edinburgh 1987), figure 11, p.43.

42 A. J. L. Winchester, *Landscape and Society*, p.52

43 Wordsworth, *Guide to the Lakes*, p.58

44 A. J. L.Winchester, *Landscape and Society*, p.89. This book is fascinating reading for anyone interested in the impact of fell farming on the Lake District landscape.

45 Angus Winchester, *The Harvest of the Hills, rural life in northern England and the Scottish Borders, 1400-1700,* (Edinburgh 2000), chapter one, 'Upland Environments and an Evolving Pastoral Economy.'

46 R. H. Lamb, *Herdwicks*, p.11

47 William Green, *The Tourist's New Guide to the Lakes,* Vol. II, (Kendal 1819), pp.265-268.

48 Winchester, *Harvest of the Hills*, p.116

49 Winchester, *Harvest of the Hills,* pp.111, 103, 119. The section on 'Theft and Violence on the Fells' includes the case of John Newcom of Rannerdale who had a history of disputes with neighbours and who killed a man with an "irnefork" in 1516 on Rannerdale Knotts.
50 A. W. Rumney, *The Dalesman* (Kendal 1936 imp.) preface and pp.70-71
51 M. L. Ryder, 'Shetland Sheep and Wool' in *The Ark*, 15 March 1982, p.99
52 See G. F. Brown, 'Introduction' to *Lakeland Shepherd's Guide*, (Workington 1985) compiled by G. F. Brown and W. Rawling, for the history of shepherd's guides and shepherds' meets.
53 See G. F. Brown and W. Rawling, *Lakeland Shepherd's Guide* (Workington 1985).
54 *Westmorland Gazette*, 23 July 1921
55 There is also a Guide for the North Pennines and the Howgill Fells. The edition published in 2000 still makes explicit the rules to be observed by the Members of the Amalgamated Association of the East, North, South and West Fells (each with their separate meets) requiring that "each member gather all Stray Sheep... upon the Moor or Common upon which his sheep usually go, four days previous to the Meeting of Exchange within his district where he resides." It contains other rules about the return of stray sheep and also about the proper marking of sheep and lambs. In addition there is a Code of Practice largely about how to operate the heafing system in a co-operative manner.
56 H. D. Rawnsley, *Months at the Lakes*, (Glasgow 1906), pp.225-227
57 H. D. Rawnsley, 'On Helvelyn with the shepherds' in *Life and Nature at the English Lakes* (Glasgow 1899) pp.158-159 and 162-163. The Helvellyn shepherds met three times a year: in July at Stybarrow Dodd; first Monday in October in Mosedale Ghyll on the road between Wallthwaite and Dockray; and there was a feast which alternated between Dockray and Thirlspot on the first Thursday after old Martinmas towards the latter part of November.
58 There is a local memory of a farmer in St. John's in the Vale in the 1930s being made to stay in his house early one morning while his neighbours gathered the fell and then counted his sheep. He had considerably more than he should have had.
59 The Eskdale May Meeting was not held in 2001, 2002 or 2003.
60 Information from Wm. Rawling, Snr. of Ennerdale, and from the late Thomas Richardson of Buttermere. See also W. R. Mitchell, *Lakeland Dalesfolk* (Clapham 1983), chapter on Herdwicks, for other examples of tups being walked on this and other routes, e.g. from West Head via Grasmere and Little Langdale. Tups were also walked from the Keswick Tup fair to Ullswater and Mardale. Mitchell's conversations with fell farmers over a fifty year period are full of authentic and accurate testimony and should not be dismissed as ephemeral journalism.
61 William Tyson, *The Herdwick Sheep,* (Keswick 1947)
62 Winchester, *Harvest of the Hills*, p.119
63 Interview with William Wilson, 2007. Wilson also pointed out that the heafing charge would come round again to form part of the pension of the tenant when he in his turn retired. There are other formulae for heafing charges based on sums of money, £10 or £12 being common.
64 William Wilberforce, *Journey to the Lake District from the Cambridge: a summer diary*, (Stocksfield 1983) p.81

Notes and References

65 William Dickinson, *The Farming of Cumberland*, (1853), p.60
66 R. H. Lamb, *Herdwicks: Past and Present*, pp.17-18.
67 William Dickinson, *The Farming of Cumberland*, (1853), pp.60-61.
68 H. A. Spedding, 'Herdwick Sheep' in *The Field*, 27 September 1873
69 William Dickinson, *The Farming of Cumberland*, (1853), pp.62-66. Dickinson also gives an outstanding example from Martinmas 1807 of a dog marking the spots where sheep were buried in snow.
70 William Dickinson, *Essay on the Agriculture of East Cumberland* (Carlisle 1853), p.41
71 People gathering at the fell have, over the years and on many occasions, broken their bones, twisted their joints and sometimes strained their hearts.
72 Wordsworth in *The Prelude* notices the dogs. He writes of a group of people being conducted from the fells on a misty summer's night and records that the "The Shepherd's Cur did to his own great joy/Unearth a hedgehog in the mountain crags/ Round which he made a barking turbulent."
73 Arthur Raistrick, *Industrial Archaeology* (1986 edn.) p.94. A variant of the sheepfold is, of course, the wash fold on the side of a beck to enable the washing of sheep a week or so prior to clipping, a practice that lasted until about the last quarter of the nineteenth century. In 1912 Garnett reported that the custom had greatly declined and that "comparatively few of the Fell and Herdwick flocks are now washed." (Garnett, p. 175)
74 Susan Johnson, 'Cumbrian Sheep that Travel' in *Cumbria,* January 1980. Mrs Johnson based her account on Hodgson's *Shepherd's Guide* of 1849.
75 Gate's *Shepherd's Guide*, 1879. Beatrix Potter did not run a fell flock from that farm, the heafed flock presumably having come to an end.
76 It was, and remains, customary and correct to walk sheep only to the far limit of their heaf before leaving them.
77 Quoted in Susan Johnson, 'Borrowdale, its land tenure and the Records of the Lawson manor' in *Transactions of the Cumberland and Westmorland Antiquarian and Archaeological Society*, Vol. LXXXI, 1981, p.68. Sea'waite, under the management of the Edmondson family since the First World War, still retains the Patrickson stock. It was in the possession of Wilson Birkett who farmed at Seathwaite by the time of the 1879 *Shepherd's Guide.*
78 *Journals of Dorothy Wordsworth*, edited by Mary Moorman (London 1971), p.50. Salving was the smearing of the sheep's skin with a greasy mixture designed to protect the sheep against ecto-parasites. On the 18 October Wordsworth worked all morning at the sheepfold "but in vain" and on the 22 October, "Wm composed without much success at the Sheepfold."
79 Wm Dickinson, *Essay on the Agriculture of West Cumberland* (1850), p.57
80 Wm Dickinson, *Essay on the Agriculture of West Cumberland* (1850), p.60
81 William Dickinson, *The Farming of Cumberland*, pp.63-65
82 Information from Syd Hardisty, Ennerdale
83 Cited in Tom Fletcher Buntin, *Life in Langdale* (Kendal 1993), pp.96-97
84 The late Thomas Richardson of Gatesgarth allowed me to consult some of his father's diaries. On the occasion referred to above, Gatesgarth had other hoggs wintering elsewhere. For the wider information see E. M. Ward, *Days in Lakeland*, pp.99-100

85 John Casson Sawrey, Copy Book, in possession of the Duddon Valley Local Historical Association.
86 William Dickinson, *Essay on the Agriculture of West Cumberland* (1850), p.59
87 Conversation in 2003 with Derwent Tyson, Threlkeld. He also told me that when he was a shepherd at Wasdale Head in the 1940s that Joe Naylor one year killed fourteen foxes.
88 Wm Dickinson, *Cumbriana* (Whitehaven 1875) pp.44-45
89 See T. W. Thompson, *Wordsworth's Hawkshead,* edited by Robert Woof, (London 1970), pp 222-223
90 William Dickinson, *Essay on the Agriculture of West Cumberland*, p.61
91 J. E. Marr, *Westmorland* (Cambridge 1909) p.87
92 Judy Taylor *Beatrix Potter, Artist, Storyteller and Countrywoman* (London 1996 edn.) pp.157-8 and *Beatrix Potter's Farming Friendship*, ed. Judy Taylor, p.25. One of my first memories of working with Herdwick sheep is the administration of carbon tetrachloride capsules at the Hollins, Ennerdale in my early teenage years.
93 William Dickinson, *Essay on the Agriculture of West Cumberland*
94 Samuel Barber, *Beneath Helvellyn's Shade: notes and sketchs in the valley of Wythburn* (London, 1892) pp.70 and 136.

Chapter 3
Herdwicks and the Lake District Landscape

1 Borrowdale is technically a shepherds' meet rather than an agricultural show. It has had very impressive turnouts of Herdwick sheep since it started in 1988. Its shepherds' meet area covers the huge majority of central Herdwick country. Its area is Borrowdale and parishes adjoining: i.e. Newlands, Buttermere, Ennerdale and Kinniside, Wasdale, Eskdale, Langdale, Grasmere and St John's, Castlerigg and Wythburn.
2 These sales did not take place in the usual way in 2001 due to Foot and Mouth Disease. The Association in that year organised a sale using a PowerPoint presentation of the available tups. In 2007, again due to FMD, the sales (with tups being present on these occasions) were not held until mid October.
3 The Ennerdale enclosure was one of the last great enclosures to occur in England: it was almost certainly an enclosure too far. The terrain has made it more or less impossible or at least prohibitively expensive to maintain the boundaries between the different allotments. Ennerdale Dale, after a flock was established there in the 1870s, was subsequently let with the flock of sheep to local farms. The bothy is now the Black Sail Youth Hostel.
4 In 1940 when the Birketts finally left Gillerthwaite they sold off 100 tups for which they achieved the very high average of £7 a head. One, 'Hop On,' made £48 and went to Wasdale Head to be shared between the flocks of Ben Ullock and Joe Naylor. Information from George Birkett, 2006.
5 H. H .Symonds, *Afforestation in the Lake District*, (London 1936), pp.35-36

Notes and References

6   HSBA Minute Book, March 1935 AGM at The Globe, Cockermouth. The resolution was moved by Frank Fawcett of Penrith and seconded by Isaac Thompson of West Head. The reference to Seathwaite is to the Duddon Valley.
7   H. H. Symonds, *Walking in the Lake District* (first published London 1933, 1942 imp.) pp.84-86
8   See H. H. Symonds, *Afforestation in the Lake District*, (London 1936), Ch. 3, 'Herdwick Sheep: the Injury to Local Farming.' On the general background see G. Berry and G. Beard, *The Lake District: a century of conservation* (Edinburgh 1980) pp.14-15, 29 and 37; also Ian Brodie, *Forestry in the Lake District*, (Friends of the Lake District, 2004)
9   H. V. Hughes, 'The Lake District; a National Park' in *Lakeland. A Playground for Britain*, ed. Katharine C. Chorley (Whitehaven, n.d.) p.25
10 G. M. Trevelyan, *English Social History*, (first published 1942, 1967 edition), p.317
11 See David Cannadine, *G. M. Trevelyan: a life in history* (London 1992) p.146
12 D. M. Matheson, *National Trust Guide: Places of Natural Beauty* (London 1950), p.15
13 Daniel Hay, *Whitehaven: an illustrated history* (Beckermet 1979 edn.) p.166; B. L. Thompson, *The Lake District and the National Trust*, p.211. The Walker family, operating as Lodore Ltd., has subsequently sold Seathwaite, Burnthwaite and Bowderdale to the Trust and still owns important fell farms in Borrowdale and Wasdale.
14 G. M. Trevelyan, 'Amenities and the State' in *Britain and the Beast,* ed. Clough Williams-Ellis (London 1938 edn.), p.183
15 Quoted in *Britain's National Parks*, (London 1959) ed. Harold M. Abrahams, p.19
16 A. Wainwright in his book *Fellwanderer* (Kendal 1966), there are no page numbers in the original.
17 David Hardman, *The History of the Holiday Fellowship*, Part One 1913-1940 (London 1981), pp.1-4
18 Trevelyan quoted in Oliver Coburn, *Youth Hostel Story* (London 1950), pp.1-2
19 C. L. Mowat, *Britain Between the Wars*, (London 1966 imp.), pp.527-528, and John Stevenson, *British Society, 1914-1945,* (Harmondsworth 1984), pp.391-393. YHA members booked over half a million nights at the hostels.
20 G. M. Trevelyan, 'Amenities and the State' in *Britain and the Beast,* ed. Clough Williams-Ellis (London 1938 edn.) pp.183-184
21 E. J. L. James, *Safeguarding Lakeland,* (Whitehaven 1928)
22 Cumbrian Regional Planning Scheme, prepared for the Cumbrian Regional Joint Advisory Committee by Patrick Abercrombie and Sydney A. Kelly, (Liverpool and London 1932) p.158. C. L. Mowat, *Britain Between the Wars,* p.465, records the percentage of insured workers unemployed in 1934 as including 36.3 in Workington and 57.0 in Maryport. J. Jewkes and A. Winterbottom in their *An Industrial Survey of Cumberland and Furness* (Manchester 1933) p.110 pointed out that of the 2,220 men insured at the Cleator Moor Employment Exchange in 1931, 1419 were unemployed – a staggering 64% rate of male unemployment.
23 Cumbrian Regional Planning Scheme, p.123

24 Cumbrian Regional Planning Scheme, p.68. It is interesting to note that the pioneering climber and mountain photographer, George D. Abraham, produced what must be the first of the modern guidebooks: on how to get around the Lakes by motor car in 1913 under the title *Motorways in Lakeland.*
25 Cumbrian Regional Planning Scheme, p.39
26 Its then land-holding was just less than 9,000 acres. It owned Wallabarrow Crag, Scafell, land at Ennerdale and Borrowdale and round Derwent Water and Ullswater: ibid p. 68.
27 Cumbrian Regional Planning Scheme, p.39
28 Kenneth Spence, 'The Lakes' in *Britain and the Beast*, p.243
29 In Buttermere, the farms of Crag House, Wilkinsyke and Rannerdale have all been acquired in the last 25 years or so.
30 E. Battrick, *Guardian of the Lakes*, pp.89-90, see also National Trust, *National Trust Properties*, March 1996 edition, p.55.
31 B. L. Thompson, *The Lake District and the National Trust* (Kendal 1946) pp.50-51
32 The Rt. Hon. Lord Birkett 'The Lake District' in *Britain's National Parks*, (London 1959) ed. Harold M. Abrahams, pp.33-36
33 Ann and Malcolm MacEwen, *Greenprints for the Countryside? The Story of Britain's National Parks* (London 1987) p.7
34 Linda Lear, *Beatrix Potter* (London 2008 imp.) p.375 and chapter 18 footnote 43 points out that Mrs Heelis tried to persuade the National Trust to establish a landlord's stock on Yew Tree, but they declined to do so.
35 See Margaret Lane, *The Tale of Beatrix Potter* (1971 imp.) Chapter 8 and W. R. Mitchell, *Beatrix Potter: her life in the Lake District,* (Settle 1998), pp.98-105.
36 Quoted in Linda Lear, *Beatrix Potter,* pp. 370-371
37 H. B. Heelis to D. M. Matheson, 31 March 1939 in *Beatrix Potter's Letters: a selection,* ed. Judy Taylor, London 1989, p.401. Tilberthwaite was part of the Monk Coniston estate which Mrs Heelis bought in 1929-1930 at a time when the National Trust could not afford it. She kept the better half of the estate and sold the rest to the Trust. She subsequently managed the whole estate for ten years.
38 Mrs H. B. Heelis to S. H. Hamer, 26 June 1926 in *Beatrix Potter's Letters.* In her will the numbers were 750 ewes, 250 gimmer twinters and 175 gimmer hoggs: see Linda Lear, *Beatrix Potter* (London 2008 imp.), p.445
39 Beatrix Potter to Eleanor Rawnsley, 24 October 1934, in *Beatrix Potter's Letters*, p.367
40 Beatrix Potter to F. Warne 16 May 1918 in *Beatrix Potter's Letters*, p.248
41 Beatrix Potter to A. McKay, 20 February 1929 in *Beatrix Potter's Letters,* p.313.
42 The marks for Hill Top were forked in both ears, H on far side. The H did not denote Heelis as is sometimes assumed, but was the mark that came with the farm when she took it over from the Hawkrigg family.
43 Sheila Richardson, *Langdale: tales from a Lakeland Valley* (Workington 1997) p.71
44 In 1928 Wedgewood was shown eleven times and won each time. Joe Cockbain was offered £50 for him but declined. In all Wedgewood won 58 first and champion prizes. See article written by R. H. Lamb to mark Joe Cockbain's retirement

in *Cumberland and Westmorland Herald*, 7 October 1933. Mrs Heelis wrote an obituary of Wedgewood which called it "the perfect type of hard, big-boned Herdwick tup... He had strength without coarseness. A noble animal." Quoted in W. R. Mitchell, *Beatrix Potter, Her Life in the Lake District* (Settle 1998) p.101

45 The original scheme stipulated that all ram lambs with potential for registration at shearling age should be seen with their mothers by the end of May in the year they were born. This was modified at the second meeting to mean that ewes fit to breed potential tups should be seen in September when the tups were being inspected for registration.

46 William Bowes recalled to me in the early 1990s that after the meeting Mrs Heelis berated the secretary, R. H. Lamb, for mentioning the ravages of the maggot fly in too much realistic detail in one of his radio broadcasts. She felt it would upset the public.

47 H. B. Heelis to J. Moscrop, April 1943, in *Beatrix Potter's Farming Friendship*, ed. Judy Taylor (London 1988) p.85

48 Information from Wm Wilson, 2007

49 George and Betty Birkett pointed out to me that it was Jerry Richardson who recommended Tommy Stoddart and his wife Florrie (née Harrison) to look after Tilberthwaite and that Mrs Heelis persuaded Bob Birkett to move from Side House to High Yewdale. She was also behind Joe Gregg's move from Taw House to Millbeck: see W. R. Mitchell, *Beatrix Potter: her life in the Lake District*, (Settle 1998) pp.101-102.

50 Beatrix Potter to A. McKay, 20 February 1929 in *Letters*, p 313. R. H. Lamb the secretary was articulate enough – being a seasoned journalist and regular broadcaster on BBC Radio. She wrote to J. Moscrop in April 1943 that "The Herdwick broadcaster R. H. L. is a less agreeable person, – at least I cannot get on with him;" Beatrix Potter's *Farming Friendship*, p.85. On Lamb's life see Geoff Brown, 'R. H. Lamb: no ordinary man of the fells' in *Cumberland and Westmorland Herald*, 9 September 1989.

51 Letter to J. Moscrop, 22 January 1930, in *Beatrix Potter's Farming Friendship: Lake District Letters to Joseph Moscrop, 1926-1943*, ed. Judy Taylor, London 1988, p.45 and Beatrix Potter's *Letters*, ed. Judy Taylor, p.297

52 See Susan Denyer, *Beatrix Potter and her farms*, (National Trust 1992)

53 Linda Lear, *Beatrix Potter* (London 2008 imp.) p.445 records that William Heelis, after his wife's death, in his own will "mandated specific numbers of heath-going sheep at Tilberthwaite as well."

54 Pearsall and Pennington, *The Lake District*, p.288

55 J. H. Cousins, *Lake District Farm Estates Ltd, A History, 1937-1977* (unpublished study for Diploma in Lake District Studies, Lancaster University) p.17, note 51. I wish to thank Jan Darrall of Friends of the Lake District for supplying a copy of this.

56 E. Battrick, *Guardian of the Lakes*, pp. 106-108. Tongue House and Long House were bought in 1983.

57 J. H. Cousins, *Lake District Farm Estates*, p.8

58 Rannerdale was subsequently sold to its tenants. It was then bought by the National Trust in 1980.

59 H. H. Symonds, *Walking in the Lake District,* 1962 edn. pp.267-8.
60 B. L. Thompson, *The Lake District and the National Trust,* (Kendal 1946), pp.35-36
61 J. H. Cousins, *Lake District Farm Estates*, p.23
62 'A Proposal to form Lake District Farm Estates Limited', signed by Norman Birkett, R. S. T. Chorley, Lord Howard of Penrith, Samuel H. Scott and H. H. Symonds, March 1937, (reproduced in Cousins, pp.30-31)
63 Elizabeth Battrick, *Guardian of the Lakes*, p.120
64 Chapel House, although occupied by its owners the Weir family, is subject to covenants made in 1937. Thorneythwaite and High Lodore remain in the ownership of Lodore Ltd.
65 'Profile of ownership in the Lake District' in *National Trust, A Vision for the Lake District after Foot and Mouth*, (Grasmere n.d., 2002?)
66 The Richardson family tenanted Gatesgarth between 1932 and 1963 at which date it was sold to them for £11,000 by Sir Claude Elliot, the Provost of Eton. The covenants, of course, remain. When in 2007-2008 the Richardson family wished to erect a sheep shed, they still needed amongst other things to get the 'go ahead' from the National Trust on top of planning permission from the National Park Authority: see *Cumbria Farmer*, January 2008.
67 There was concern for instance in the mid 1990s when the landlord's stock at Kidbeck in Nether Wasdale was deliberately dispersed by the Trust at the end of a tenancy – presumably to relieve grazing pressure on Nether Wasdale common, something that was in fact dealt with structurally a few years later through the entry of the common into an ESA agreement. There was also concern around 2000 when the Trust bought grazing rights at Braithwaite, presumably, it was felt, to take them out of use.
68 Bruce Thompson, *The Lake District and the National Trust* (Kendal 1946) p.31
69 H. H. Symonds, *Afforestation in the Lake District,* (London 1936) pp.28-29
70 H. H. Symonds, *Afforestation in the Lake District*, p.67
71 J. de Vasconcellos, *She Was Loved. Memories of Beatrix Potter* (Kendal 2003), pp.37-42
72 *Beatrix Potter's Letters*, p.366. Elizabeth Battrick in *Guardian of the Lakes*, writes about restrictive covenants as follows "landowners wishing to protect the amenities of their land without parting with possession have in the past and can still enter into covenants with the Trust... covenants have to be negative in substance (since a positive covenant would not run with the land so as to be binding on successive owners), but by their use land can be protected against, for example, building development, quarrying, afforestation and camping and caravanning." It has to be said of course that the planning system probably gives adequate safeguards against all these things – and that covenants do not of themselves protect fell farming.
73 Linda Lear, *Beatrix Potter,* (London 2008 imp.), p.445
74 B. L. Thompson, *The Lake District and the National Trust* (Kendal 1946), pp.50-51
75 Lake District Park, *Commons Project, 1988-1990*

Chapter 4
Recent Years: Challenges and Changes

1. HSBA Minute Book, 22 May 1950
2. See W. R. Coulthard on the Mungrisdale Swaledale Sheep Show (which was founded in 1929) in *Swaledale Millennium Journal*, (Penrith 2000) p.87 where he writes that prior to 1929 there had been the old Haltcliffe Ram Show which was mainly for Herdwicks.
3. Hill Farm Research, 2nd Report, HMSO 1953, cited in Andrew Humphries, *Hill Farming in the Cumbria Uplands: a Cultural Perspective in the 1990s* (Newton Rigg, 1994), Appendix 2.
4. Margaret Capstick, *Some Aspects of the Economic Effects of Tourism in the Westmorland Lake District,* (University of Lancaster, Department of Economics, 1972), p.52 and table 4.1a.
5. 'Dick's Permission' was unbeaten in its class from 1953 to 1957. It was also honoured by the National Sheep Association on the basis of its record as the most outstanding sheep in the British Isles. Derwent Tyson had an outstanding achievement when at the Fell Dales Association 100th Anniversary Show in 1964 he won both the male and the female championship.
6. Personal observation: lambs betraying a bit of Herdwick were reckoned by the shepherds to be hardier at birth than the Swaledales.
7. *Lake District Herald*, 14 October 1989
8. BBC News website for 4 March 2003
9. The HSBA from the late 1990s encouraged members to carry out scrapie genotyping: a process which led to the culling of large numbers of tups and which remains controversial.
10. Geoffrey Berry, *A Tale of Two Lakes,* (Kendal 1982), pp 58-60. Mireside also had the fell land called The Side on the other side of Ennerdale Water, but there were no extra inbye resources with it and Mireside like many another fell farm had to acquire land away from the holding to manage the increased stock it had.
11. Technical advances helped: by the late 1980s big bale silage was common, ensuring a guarantee of some decent winter feed and All Terrain Vehicles were developed which considerably lengthened the working life of shepherds as well as enabling access with supplementary feed to some (though definitely not all) remote places.
12. Ann and Malcolm MacEwen, *National Parks: conservation or cosmetics* (London 1982), pp.203, 211
13. *Choices for Farmers, Lake District National Park,* GCSE Resource Guide 3, prepared by Andrew Humphries (1989). Figures up-dated in March 1991 by G. F. Brown.
14. W. H. Pearsall and W. Pennington, *The Lake District* (first published 1973, 1989 edn.) p.177 and p.182
15. Derek Ratcliffe, *Lakeland: the Wildlife of Cumbria* (London 2002), pp 316-317
16. Jeremy Hunt, 'Hardy Herdwick 'natural survivor" in *Farmers' Weekly*, 15 February 2008
17. Cumbria Foot and Mouth Disease Inquiry Report, (Cumbria County Council) September 2002, p.9

18 Scrapie resistance testing in Herdwick sheep began in the late 1990s and involved substantial culling of low resistance tups. Many feel that it has weakened the quality and the genetic diversity within the breed.
19 An account which gives an authentic impression of the feelings and fears involved can be found in *To Bid Them Farewell*, (Kirkby Stephen 2004) by Cockermouth auctioneer Adam Day who valued many Herdwick sheep before culls.
20 MacEwen, *National Parks*, p.205
21 Cumbria Foot and Mouth Disease Inquiry Report, September 2002, p.90
22 Natural England, *Strategic Direction, 2006-2009*
23 The Trust claims a landlord's stock of 22,000 sheep. But it has to be pointed out that not all these are Herdwicks, and those that are designated as Herdwick are on occasion cross-bred rather than pure: witness the difficulties sometimes experienced at sheep livering days when Herdwicks are meant to pass from one tenant to the next.
24 For instance, by campaigning for vaccination rather than culling: this was the message of a letter to *The Times* of 5 April 2001 signed by Fiona Reynolds of the National Trust, Jonathan Dimbleby of the Soil Association, Simon Lyster of the Wildlife Trusts, Graham Wynne of the RSPB and Gilbert Tyson, chairman of the Herdwick Sheep Breeders' Association.
25 One imagines that Susan Denyer and Nick Hill were influential amongst the writers of the vision document. There has been progress on these and other small schemes since 2001. They have been carried out using a variety of resources including those of the Fells and Dales LEADER + Programme.
26 *State of Nature: the upland challenge*, (English Nature 2001), p.53
27 *State of Nature: the upland challenge*, pp.5-6
28 *State of Nature: the upland challenge*, pp.52-55 and p.62

Chapter 5
The Way Forward for Herdwick Sheep

1 Report of the Working Party on Grazing in the New Forest, September 1991 (the Illingworth Report), p.53
2 UK National Action Plan on Farm Animal Genetic Resources, report to Defra and the Devolved Administrations, November 2006, p.52. The Sheep Trust has also more recently gained acceptance of the need for additional measures where sheep breeds are highly geographically concentrated.
3 *UK National Action Plan on Farm Animal Genetic Resources, 2006*, Recommended Action 29 suggests that although the National Co-ordinator should intervene in the discussion about resources in the new rural development funding regime, this is on the basis of "other than headage payments". The present writer put some proposals to Defra and English Nature on the use of the native breeds provision immediately post-FMD but elicited no positive response. Headage payments for breeds at risk would, however, be the most direct and cost effective way of dealing with the issue.
4 Ibid., p.13
5 Philip Lowe, 'After Foot and Mouth' in *Re-Making the Landscape*, ed J. Jenkins (London 2002) p.173
6 J. D. Wood and others, *An Investigation of Flavours of Meat from Sheep Grown*

Slowly or More Quickly on Grass Diets (MAFF LS1904), University of Bristol. Research project conducted for Ministry of Agriculture, Fisheries and Food, October 1996-June 1997

7 W. J. Onions, *Wool: an introduction to its properties, varieties, uses and production* (London 1962) p.159. Onions gives a figure of 310,000lbs for the Herdwick clip in 1959.

8 See, for instance, article on Vic Gregg and Millbeck, Great Langdale in *Farmers' Weekly*, 2 March 1984

9 As at June 2006, there was in stock 57,359 kilos of light Herdwick wool and 32,400 kilos of dark Herdwick wool. The BMWB price per kilo for light was £0.10 and for dark was £0.08. The National Trust scheme paid £0.37 for light and £0.24 for dark. 114 farmers were involved in the scheme receiving per farm an average of £245. Information from David Townsend, who ran the scheme.

10 'Herdwick carpet fails to save jobs' in *Westmorland Gazette*, 18 November 2005

11 Memorandum submitted by Geoff Brown to Select Committee of Environment, Food and Rural Affairs, 1 April 2003. The Wool Clip co-operative of women interested in the local wool economy has developed an outstanding national festival of all things woollen, the Wool Fest.

12 Bob Orrell has criticised the Association on various occasions over the last ten or so years on the grounds that not enough promotion is done to establish the breed more widely. The present writer continues to think this is unrealistic and that it can detract from the main message of Herdwick sheep which is (as Orrell himself has very effectively pointed out on more than one occasion) their role in the particular and exceptional landscape of the Lake District.

13 Other breeders of large annual consignments of, say, ten or more tups in recent years are the Harrisons of Butterilket, the Dickinsons of Brockstones, Kevin Wrathall at Cockley Beck, the Naylors at Bowderdale, Gowan Grave of Gosforth, Jean Wilson of Dockray, the Wilkinsons at Tilberthwaite, Joe Folder of Cockermouth and the Nicholsons of Lorton.

14 There is a serious imperative of avoiding what has happened to another of Cumbria's native animals, the Fell Pony. There are now only nine fell-going herds of Fell Ponies in Cumbria, in spite of massive interest in the breed from all over the country and beyond: *Cumberland and Westmorland Herald,* 19 April 2008

15 If adopted this would be similar to the situation with Shetland Sheep, where there is a Flock Book Society for Shetland sheep in the Shetland Isles and a Shetland Sheep Society which operates under a strapline of "caring for Shetland Sheep outside the Shetland Isles."

Chapter 6
Herdwick Country

1 National Trust, *Economic Benefit of the National Trust's Work in Cumbria,* report by SQW, 2000, p.4

2 See for instance *Westmorland Gazette*, 11 February 2005 which contains letters of opposition from local hill farmers, residents, the chairman of the Beatrix Potter Society and the secretary of the Herdwick Sheep Breeders' Association. The following week's paper contained a lengthy statement from John Darlington, the area manager for the Trust.

3. 'National Trust told not to subsidise hill farmers' in *Cumberland and Westmorland Herald*, 26 November 2005
4. 'Tenant farms under threat due to shortage of suitable farmers' in *Cumberland News*, 23 February 2007
5. 'National Trust offers part-time agreements' in *Farmer's Weekly*, 10-16 September 2004
6. Quoted in A. and M. MacEwen, *Greenprints for the Countryside: the story of Britain's National Parks* (London 1987), p.48
7. MacEwen, *Greenprints for the Countryside*, pp.131-133
8. Lake District National Park, *Management Plan 2003*, section 4.3
9. *Cumberland News*, 30 March 2007. The Authority also voted to sell off Wood End which has land going on to the western shore of Bassenthwaite lake when it becomes vacant.
10. G. Chitty, *Proposed Lake District World Heritage Site: study of cultural landscape significance,* November 2002, p.30 and p.22
11. Susan Denyer, 'The Lake District: a proposed World Heritage Cultural Landscape Site' in *The Cultural Landscape* (ICOMOS-UK 2001) pp.132-133 and p.139
12. Peter Fowler, 'Cultural Landscape: Great Concept, Pity About the Phrase' in *The Cultural Landscape*, (ICOMOS-UK 2001), p.71
13. Some of this work has been done by a Fells and Dales LEADER + project, Hill Farming Systems. See www.cumbriahillfarmscharter.org.uk. Cumbria County Council owns the Herdwick farm, Middle Row, Threlkeld, located next to the Blencathra Field Studies Centre and the Forestry Commission owns what is left of Gillerthwaite in Ennerdale.
14. Work is being done on some of these issues through a Carnegie UK Trust project.
15. On 'ecomuseums' see P. Davis, *Ecomuseums: A Sense of Place* (Leicester UP 1999).
16. Institute for European Environmental Policy, *Final Report for the Study of HNV Indicators for Evaluation Report for DG Agriculture,* October 2007, p.16
17. IEEP *Final Report*, p.12. Another avenue would be to explore UNESCO Biosphere Reserve status.
18. Susan Denyer, 'The Lake District: a proposed World Heritage Cultural Landscape Site,' p.140.

Index of People, Practices and Organisations

Abbott, Wm	19, 24, 28
Abercrombie, Patrick	79
Agriculture, Board of	10
Afforestation	73-6, 98-9
Allison, JB	29
Balliol College	89
Bailey and Culley	10-13
Bainbridge, Thos	39
Banner, Delmar	99
Bell, J	28
Bellas, Joseph	35
Bindloss, EW	31
Birkett, Betty	93
Birkett, George	34
Birkett, J (How Hall)	25
Birkett, W (Gillerthwaite)	25, 27
Birkett, John and Ruth	137
Birkett, Rt Hon Norman	87
Black-faced sheep	10-11, 15-17
Bland, Wm (secretary of HSBA)	108-9
Bland family, West Head	32, 133
Blowfly	50-1, 91
Blue tongue virus	152
Bluefaced Leicester (sheep)	105
Bowman, John	14
Bowes, Thos	30
Bowes, William	106
Bowles, Dianna	126
Briggs, Robert	14
Brown, W	31
Brownrigg, Sam	30
Brownrigg, Ernest	66
Buntin, JF	39
BSE	110
Carpets	131-2
Carnegie UK Trust	141
Chernobyl nuclear accident	109
Chorley, RST (Lord)	94-5
Clarke, James	7, 11
Clemitson, T	29
Clipping Days	49-51
Cockbain, J	28
Cockbain, Mrs	28, 31, 64
Common land	22, 54-5, 101, 113, 116, 123, 147, 149
Conservation Grazing Schemes	124-6
Co-operative Holidays Association	78
Council for the Preservation of Rural England (CPRE)	78-9
Countryside and Rights of Way (CROW) Act	116
Countryside Agency	115
Countryside Commission	143
Covenants to protect landscape	86, 97, 99
Cowx, J	29, 31, 64
Cowman, Joe (secretary of HSBA)	107-8
Cumbria Farmer Network	150
Dawson, H	26
Defra	113, 121
Denton, Thos	43, 53
Denyer, Susan	147-8, 152
Devon, R	26
Dodgson, S. D. Stanley	13, 22-3, 26, 29, 37
Dickinson, Wm	15, 17, 18, 63-5, 67, 69-70
Dogs	64-5, 148
Dunn, George	139-140
Dunning, John	143
Edmondson family	27, 30
Ellwood, Rev Thos	9, 17
Enclosure	22, 53
English Nature	117, 122
Environmentally Sensitive Area (ESA) scheme	113
Ewe Inspection	37
Essays on Herdwick	18-9
Fawcett, Frank	37-8
Fell Dales Association for the Improvement of Herdwick Sheep	16, 25, 72
Fell and Rock Climbing Club	76-77

175

Fleming, Sir Daniel	14, 44	Kinsman, David	9-10
Flockbook of Herdwick Sheep	22, 37	Lake District Farm Estates	86, 94-5
Fluke, Liver	70	Lamb, R H (secretary of HSBA)	10, 23, 37, 39, 41, 48, 54, 63, 75
Forestry Commission	73-74, 89, 94, 96	Lamplugh, Sir J	43
Foot and Mouth Disease (FMD)	103, 114, 115-120, 123	Land Futures	145-6
Fox, Wilson	48	Landlords' flocks	43-8, 88, 136
Fox hunting	70-71	Lawrence, WT	52
Fountains Abbey	53	LEADER programme	141, 150
Furness Abbey	42, 53	Leck, Mrs	28
Friends of the Lake District	74-6, 94-5	Leconfield, Lord	28, 64
Garnett, FW	10, 12, 17, 23, 25, 55	Lleyn, (sheep)	127
Genetic diversity	126-8, 149	Lonsdale, Earl of	27, 73
Grave, Simeon	28 (illus), 30	Lowther, Sir J	44
Grave, Gowan	106-7	MacEwen, A and M	86
Green, Wm	54	Maggots, see blowfly	
Gregg, J N (ewe inspector)	25, 27, 38	Manchester Corporation	89, 97
Gregg, J V	69, 106	Marshall, WH	19, 86
Haltcliffe Ram Show	103	Martin, George	54
Hardisty, Harry	106	Maurice, Oliver	119
Hardisty, Syd	106	Meat	128-30
Harrison, Joseph	48, 66-7	Metcalfe, John	141-2
Harryman, J	26 (illus)	Ministry of Agriculture	103, 112
Hartley, Gilbert	107	Moscrop, Joseph	91, 93
Hartley, Tyson	107	'Mule' sheep	105
Hartley, Anthony	114, 129, 133	Mutton	22-3
Hawell, Joseph and Edward	16, 29, 64	National Land Fund	95-6
Hawkrigg, William	66	National Sheep Association	107
Heafing amongst Herdwick sheep	42, 43, 62-3, 114, 116	National Parks (especially Lake District)	85-7, 99-102, 111, 142-6
Heelis, Mrs H B or Mrs Wm (Beatrix Potter)	39, 73, 87-94, 99-100	National Trust	23, 47, 76-7, 86, 95-7, 101-2, 109, 118-21, 136-42
Heelis, Wm	87, 91-2	Natural England	118
Herdwick Sheep Breeders' Association	10, 22, 24, 25-31, 32, 35-39, 40-1, 51, 72-3, 75, 91-2, 99, 109, 132-3, 141	Naylor family	31, 33
		Nelson, Edward (Snr and Jnr)	19, 20, 21, 25, 27
Hill Farming Act, 1946	40, 103	Nixon, Peter	128
Hill Livestock Compensatory Allowance	111	Over-grazing	112, 116-8, 123
		Open Spaces Society	138-9
Hinde, Harry	106	Palmer, WT	50-1
'Homing Instinct' amongst Herdwick sheep	63	Pape, DN	23
		Park, WN	25
Humphries, Andrew	20	Patrickson sheepfold	67
Jackson, John	14	Pearson, Allan	28
Jackson, Wm	43-4	Pearson, Thos	14
Johnson, Neil	139	Pigou, Prof A C	86
Johnson, Mrs Susan	66, 95	Planning, Town and Country	79-87
		Plaskett, J (ewe inspector)	25, 39

Index

Ponsonby family, Kinniside 30
Potter, Beatrix: see Heelis, Mrs H B
Ratcliffe, DA 112
Rawnsley, Canon HD 16, 23-4, 51, 99
Rawnsley, Eleanor 89, 99
Rawnsley, Noel 23
Rawling family, (Ennerdale) 30, 37-8, 41
Rawling, Jonathon 37-8, 41
Rawling, Wm (ewe inspector) 38 (illus), 41
Rawling, Thos 27
Rawling, William Snr 108
Rawlinson, Robert 43
Richardson family, (Watendlath) 133
Richardson, Betty (see also Birkett) 35
Richardson, Isaac 35
Richardson, Jerry 25, 34, 69, 92
Richardson, John Snr. 34
Richardson, Jos 35
Richardson, Johnny 34-5
Richardson, Simpson 25, 27, 35
Richardson, Joseph (Ennerdale 34
Richardson, Willie 35
Richardson, Wm (Thwaites) 34
Richardson, Thos 34
Ridley, Thos 27
Rigg, Wm 37
Robinson, Harry 106
Rogers, Joseph 54
Roper, J 25, 31
Rothery, J 25, 28
Rough Fell sheep 18, 115, 126, 149
Rowlandson, Thos 14
Royal Show 16, 20, 21, 23, 107
Rumney, A W 55
Rural Development Programme for England 122-3
Ryder, M L 10
Sanderson, J 28
Sawrey, John 52-3, 69
Scandinavians 9-10
Sharp, Andrew 129
Sheep folds 66-7
Sheep scoring numerals 25
Shepherd's Guides 18, 21, 56-61, 90
Shepherds' Meets 57-61, 110
Shetland sheep 9-10, 18
Sites of Special Scientific Interest (SSSI) 117-8
Snow 67-8
Spaelsau sheep 10
Spedding, H A 64
Spence, K 85
Stable, Wilson 54
Stoddart, family 88
Storey, Thos 88
Swaledale sheep 18, 40-1, 103-5, 108
Symonds, H H 67, 73-6, 94-5, 97-9
Task Force for the Hills 122
Tenancies 88
Todhunter, J 29, 64
Thompson, Isaac 25, 27, 51
Thompson, BL 85, 95, 97
Thwaites, Thos 31
Towers, Thos 54
Trevelyan, GM 73, 76-79
Troughton family, Thwaites 30
Tup registration 36-7, 72, 92, 133
Tyson, Derwent 106
Tyson, Ernest 106
Tyson, Gilbert (chairman of HSBA) 106
Tyson, Teddy (chairman of HSBA) 106-7
Tyson, Wm (secretary of HSBA) 26, 62, 107
Tyson, Wm (Farthwaite) 29, 32
Tyson, Wm (18th century, Muncaster) 13
Tyson, Wm (Black Hall) 54
Tyson, J (Ennerdale) 14
Tyson, J (Eskdale) 14
United Utilities 97
Upland Entry Level Stewardship (UELS) 114
Upland Management schemes 143
Viewing Days, Sheep 47-8
Vikings 8, 9
Wainwright, Alfred 78
Walker, Herbert 77
Webster, Crayston 22
Weir family 31, 106
West Cumberland Fell Dales Association 14, 16
Wethers 28, 48-9, 52
Whyte, Ian D 8-9
Wilberforce, William 63
Williamson, J 25

Wilson, Christopher 16
Wilson, Dick 33, 106
Wilson, George (chairman of HSBA) 107
Wilson, Jean 106
Wilson, John (Keskadale) 19, 24
Wilson, John (Wasdale) 15 (illus)
Wilson, Harry 107
Wilson, R M 25, 27, 31-2
Wilson, William (Herdwick Billy), secretary of HSBA 26, 27, 32-3, 92
Wilson, Wm Jnr 33
Winchester, Angus 53-5
Wintering 68, 69, 103
Wool 19, 48-52, 108, 130-2
Wordsworth, William 7, 53, 67, 85, 148
Wordsworth, Dorothy 67
World Heritage Site 119, 146-152
Youatt, Wm 12
Youth Hostels Association 78

Index of Farms

Ashness, Borrowdale	33, 96		54, 69, 86, 133
Baskell, Ulpha	116	Gill, Kinniside	32
Beckside, Sandwick	101, 144-5	Gill, Nether Wasdale	95
Birkhow, Little Langdale	77, 136	Gill Bank, Eskdale	95
Birkrigg, Newlands	34	Gillbrow, Newlands	31
Black Hall, Ulpha	27, 54, 74, 94, 96, 98, 115	Gillerthwaite, Ennerdale	14, 27, 34-5, 50, 61, 74
Black How, Cleator	41	Glencoyne, Glenridding	31, 33, 45, 52, 54, 96, 107
Black Sail	73		
Bowderdale, Wasdale	96, 136	Godferhead, Loweswater	41
Bram Crag, St Johns	26	Grassguards, Ulpha	52, 69
Bridge End, Lorton	28	Greenrigg, Caldbeck	64
Bridge End, Thirlmere	59	Greystone House, Thwaites	31
Brighouse, Ulpha	94	Harrowhead, Nether Wasdale	95
Brimmer Head, Grasmere	96, 136	Harry Place, Gt Langdale	77
Broadrayne, Grasmere	107	Hartsop Hall, Hartsop	95
Brotherilkeld or Butterilket	25, 31, 40, 48, 53-4, 66-7, 74, 94, 96-8	Hazel Head, Ulpha	43, 94, 116
		High Lodore, Borrowdale	35
Browside, Duddon	94	High Row, Threlkeld	28, 64
Buckbarrow, Nether Wasdale	95	Hill Top, Sawrey	90-2
Burnthwaite, Wasdale Head	77, 139	High Nook, Loweswater	95
Causeway Foot, Keswick	28, 64	Hollins, Ennerdale	30, 73
Chapel Hill, Mardale	51	Howe Green, Hartsop	95
Chapel House, Borrowdale	19	Howe Head, Coniston	92
Chapel House, Uldale	64	Hoathwaite, Torver	96, 150
Cockley Beck, Duddon	77	Hollows, Borrowdale	96
Crag, Ennerdale	25	Howes, Eskdale	110
Crag House, Buttermere	96	How Hall, Ennerdale	14, 25, 96
Dale End, Little Langdale	59, 77	Kentmere Hall	16, 18, 29, 44
Darling How, Lorton	92	Keskadale, Newlands	19, 26, 40
Dowthwaite Head, Matterdale	68	Kidbeck, Nether Wasdale	136
Dungeon Ghyll, Gt Langdale	77	Kiln How, Borrowdale	35
Farthwaite, Kinniside	32	Kirkstile, Loweswater	19
Fell Foot, Little Langdale	77, 96	Lanthwaite Green	27
Fellside, Bootle	34	Little Braithwaite, Newlands	41
Fenwick, Thwaites	30, 49, 96, 136	Longlands, Uldale	64
Fold Head, Watendlath	136	Longthwaite, Borrowdale	95
Folds, Ulpha	30	Lonscale, Keswick	16, 29
Forest Hall, Selside	54	Low Nest, Keswick	29, 64
Foulsyke, Loweswater	19	Low Snab, Newlands	35
Gatesgarth, Buttermere	13, 25, 35, 53,	Middlefell, Gt Langdale	47, 77

Middle Row, Threlkeld	35
Middle Row, Wasdale	27, 31, 33, 95, 130
Milkingstead, Eskdale	142
Millbeck, Gt Langdale	69, 129
Millbeck Hall, Keswick	30
Mirehouse, Keswick	29, 64, 95
Mireside, Ennerdale	14, 95, 136
Monk Coniston estate	93
Naddle, Haweswater	22, 108
Nook, Rosthwaite	32, 96
Oaks, Loughrigg	29
Patterdale Hall	19, 28, 54
Penfold, Dockray	31
Penny Hill, Eskdale	29, 94
Pikeside, Ulpha	25, 34-5, 46, 94
Rannerdale, Buttermere	95
Robin Ghyll, Gt Langdale	77
Robinson Place, Gt Langdale	77
Routen, Ennerdale	25
Row Head, Wasdale	96
Rydal Hall	14
Sand Ground, Hawkshead	45
Seathwaite, Borrowdale	27, 30, 34, 63, 77, 96, 136
Seatoller, Borrowdale	47, 96
Shap Abbey	54
Side, Great Langdale	77
Simon Keld, Kinniside	110
Skelgill, Newlands	28, 30
Skiddaw Forest	64, 112
Souterfell, Mungrisedale	35
Standing Stones, Kinniside	32
Stoneycroft, Newlands	33
Stool End, Gt Langdale	77
Stonethwaite, Borrowdale	47, 53
Swinside, Ennerdale	25, 27, 34, 66
Swinside, Thwaites	34
Swinside End, Kinniside	14
Taw House(Tows), Eskdale	14, 27, 35, 54, 96
Thirlspot, Thirlmere	59
Thornhome, Calder Bridge	35
Thornthwaite Hall, Bampton	21, 54
Thrang, Duddon	94
Thwaite Yeat, Thwaites	30
Tilberthwaite, Coniston	35, 54, 68, 88, 94, 137
Townhead, Grasmere	96
Troutal, Seathwaite	96
Troutbeck Park	28, 88, 91, 93-4
Turner Hall, Seathwaite	27, 30, 97, 107, 133
Wallabarrow, Duddon	96, 119
Wall End, Gt Langdale	77
Wasdale Head Hall	95, 129
Watendlath, Borrowdale	26-7, 33, 70, 133
West Head, Thirlmere	27, 51, 66, 97, 133
Wha House, Eskdale	96, 136
Wilkinsyke, Buttermere	96
Woodend, Ulpha	108
Woodhall, Hesket New Mkt	27
Wood How, Nether Wasdale	33, 149
Woolpack Inn, Eskdale	16
Yewdale, Coniston	94, 121, 136-42
Yew Tree, Coniston	39, 89, 129
Yew Tree, Rosthwaite	47, 95, 96, 129

Photographs

Photographs are from the following sources:

1) the Herdwick Sheep Breeders' Association Flock Books;
2) courtesy of the families whose members they concern – especially Messrs Birkett, Bowes, Grave, Harrison, Harryman, Rawling, Richardson, Tyson, Wilkinson, Wilson;
3) Bruce Wilson, Swarthmoor;
4) Ian Brodie.

Thanks again to Louise Rawling, Ennerdale and Dorothy Wilkinson, Tilberthwaite for their help in sourcing photographs.